SCIENTIFIC STRATEGIES TO SAVE YOUR LIFE

A Statistical Approach to Primary Prevention

STATISTICS: Textbooks and Monographs

Volume 1: The Generalized Jackknife Statistic, *H. L. Gray and W. R. Schucany*

Volume 2: Multivariate Analysis, *Anant M. Kshirsagar*

Volume 3: Statistics and Society, *Walter T. Federer*

Volume 4: Multivariate Analysis: A Selected and Abstracted Bibliography, 1957-1972, *Kocherlakota Subrahmaniam and Kathleen Subrahmaniam* (out of print)

Volume 5: Design of Experiments: A Realistic Approach, *Virgil L. Anderson and Robert A. McLean*

Volume 6: Statistical and Mathematical Aspects of Pollution Problems, *John W. Pratt*

Volume 7: Introduction to Probability and Statistics (in two parts)
Part I: Probability; Part II: Statistics, *Narayan C. Giri*

Volume 8: Statistical Theory of the Analysis of Experimental Designs, *J. Ogawa*

Volume 9: Statistical Techniques in Simulation (in two parts), *Jack P. C. Kleijnen*

Volume 10: Data Quality Control and Editing, *Joseph I. Naus*

Volume 11: Cost of Living Index Numbers: Practice, Precision, and Theory, *Kali S. Banerjee*

Volume 12: Weighing Designs: For Chemistry, Medicine, Economics, Operations Research, Statistics, *Kali S. Banerjee*

Volume 13: The Search for Oil: Some Statistical Methods and Techniques, *edited by D. B. Owen*

Volume 14: Sample Size Choice: Charts for Experiments with Linear Models, *Robert E. Odeh and Martin Fox*

Volume 15: Statistical Methods for Engineers and Scientists, *Robert M. Bethea, Benjamin S. Duran, and Thomas L. Boullion*

Volume 16: Statistical Quality Control Methods, *Irving W. Burr*

Volume 17: On the History of Statistics and Probability, *edited by D. B. Owen*

Volume 18: Econometrics, *Peter Schmidt*

Volume 19: Sufficient Statistics: Selected Contributions, *Vasant S. Huzurbazar (edited by Anant M. Kshirsagar)*

Volume 20: Handbook of Statistical Distributions, *Jagdish K. Patel, C. H. Kapadia, and D. B. Owen*

Volume 21: Case Studies in Sample Design, *A. C. Rosander*

Volume 22: Pocket Book of Statistical Tables, *compiled by R. E. Odeh, D. B. Owen, Z. W. Birnbaum, and L. Fisher*

Volume 23: The Information in Contingency Tables, *D. V. Gokhale and Solomon Kullback*

Volume 24: Statistical Analysis of Reliability and Life-Testing Models: Theory and Methods, *Lee J. Bain*

Volume 25: Elementary Statistical Quality Control, *Irving W. Burr*

Volume 26: An Introduction to Probability and Statistics Using BASIC, *Richard A. Groeneveld*

Volume 27: Basic Applied Statistics, *B. L. Raktoe and J. J. Hubert*

Volume 28: A Primer in Probability, *Kathleen Subrahmaniam*

Volume 29: Random Processes: A First Look, *R. Syski*

Volume 30: Regression Methods: A Tool for Data Analysis, *Rudolf J. Freund and Paul D. Minton*

Volume 31: Randomization Tests, *Eugene S. Edgington*

Volume 32: Tables for Normal Tolerance Limits, Sampling Plans, and Screening, *Robert E. Odeh and D. B. Owen*

Volume 33: Statistical Computing, *William J. Kennedy, Jr. and James E. Gentle*

Volume 34: Regression Analysis and Its Application: A Data-Oriented Approach, *Richard F. Gunst and Robert L. Mason*

Volume 35: Scientific Strategies to Save Your Life: A Statistical Approach to Primary Prevention, *Irwin D. J. Bross*

OTHER VOLUMES IN PREPARATION

SCIENTIFIC STRATEGIES TO SAVE YOUR LIFE

A Statistical Approach to Primary Prevention

IRWIN D. J. BROSS

MARCEL DEKKER, INC. New York and Basel

Library of Congress Cataloging in Publication Data

Bross, Irwin D. J.
Scientific strategies to save your life.

(Statistics, textbooks, and monographs ; v. 35)
Bibliography: p.
Includes index.
1. Medical statistics. 2. Epidemiology--Statistical methods. 3. Public health. I. Title. [DNLM: 1. Biometry. 2. Science. WA 950 B874s]
RA409.B837 614.4'072 81-504
ISBN 0-8247-1273-0 AACR1

MARCEL DEKKER, INC.

270 Madison Avenue, New York, New York 10016

Current printing (last digit)
10 9 8 7 6 5 4 3 2 1

PRINTED IN THE UNITED STATES OF AMERICA

CONTENTS

ACKNOWLEDGMENT

This book is dedicated to the unsung heroines who have been the most effective defenders of the public health and safety against the misuse and abuse of technologies. Several of the heroines I have met have carried on this struggle while they themselves were dying of an incurable disease. When I speak of "concerned citizens" I usually have in mind these dedicated women.

There are two women whose patience, hard work, and steadfastness have made this book an actuality despite many discouragements. They are my wife, Rida Bross, and my secretary, Mary Ann Kuczmarski.

SCIENTIFIC STRATEGIES TO SAVE YOUR LIFE

A Statistical Approach to Primary Prevention

chapter 1

YOUR LIFE AND SOME "FACTS OF LIFE"

Three steps to save your life: Understanding the action situation, getting involved, and knowing where to start.

STEPS TO SAVE YOUR LIFE

It is my job to try to save your life and I take my job seriously.

Allow me to briefly introduce myself and my work. After getting my doctorate in Experimental Statistics from the University of North Carolina in 1949, I followed my major professor, W. G. Cochran, to the School of Hygiene and Public Health at the Johns Hopkins University in Baltimore. Here I served my apprenticeship in biostatistics and public health and worked with C. Windsor, M. Merrell, A. Lilienfeld, and others. In 1952, I moved to Cornell University Medical College and the Sloan-Kettering Institute for Cancer Research in New York City, where I became directly involved in epidemiology and cancer research. With E. Wynder and others, I got into the cigarettes-and-lung-cancer and other major public health issues and began to learn some of the "facts of life" about these controversies. In 1959 I went to the Roswell Park Memorial Institute for Cancer Research in Buffalo, New York, where I became a state health bureaucrat with some direct responsibility for protecting the health and safety of the public that was paying my salary. So saving lives is my business, and I am an old pro with more than 25 years of experience in this business.

My primary purpose in writing this book is to help you to save your life and to save as many other lives as I possibly can.

It isn't easy to save lives, but the difficulties are not what you might think they are. Discovering what is producing deaths or disabilities--identifying the hazards--is a difficult task, but not the most difficult task. During the 25 and more years that I've been doing public health research, I have made at least half a dozen major scientific discoveries with a substantial and immediate potential for saving hundreds or thousands of lives. These findings have been made in collaboration with top epidemiologists and cancer scientists or with members of my own staff. In addition to the major discoveries, our research has turned up more than a score of minor ones. Unfortunately, in the past generation there has been a great opportunity for scientists to discover new health hazards because so many have been added to the environment in these years. If you have the competent people, the necessary resources, and the know-how to do the job, it is generally possible with a bit of hard work to find out just what is killing people or disabling them.

The hard part is trying to stop this unnecessary death and disability. Once you know what is producing a hazard, it will usually be possible to use this scientific knowledge to devise effective public health measures. However, getting the effective public action that is necessary to actually save lives is an extremely slow, difficult, and frustrating task. In my many years of experience with this task, I have learned some hard lessons which I would call the "facts of life." My purpose here is to pass on this experience to you.

Why is it so hard to get the public action that is needed to save your life? Why do the controversies on health issues drag on interminably? In a word, the answer is: Money. Nearly all of the major new health hazards that have come into our environment in recent years are in some way connected with the misuse or abuse of products and processes of modern technology. They have come into the environment because those who were promoting this technology had something to gain. Any restriction on the foolish, unnecessary, or dangerous uses of this technology will be bitterly resisted by those with a stake in the technology who feel they have something to lose. The stakes are generally high in these health issues--many millions or billions of dollars may be involved.

In every issue that I have been involved with, I have heard the defenders of the technologies piously proclaim that, if anyone were *really* hurt or killed, they would be the first to act. These same people then refuse to even consider the scientific facts that human beings *are* being hurt or killed. When the chips are down, they act in their own self-interest and are indifferent to what happens to other human beings. This is why it is so difficult to take the steps necessary to save your life even when we know what steps to take.

STEP 1: UNDERSTANDING THE ACTION SITUATION

It is easy, of course, to blame someone else for a public health problem--but this doesn't solve the problem. I don't condone the way that those with economic interests in hazardous technology have acted to protect those interests irrespective of what they are doing to the health and safety of other human beings. Rather, I view this as one of the hard "facts of life" that anyone who is trying to protect the public will have to take into account in his or her planning. My object here is to give a realistic picture of what I will call the "action situation" in public health issues--neither an over-optimistic and idealistic picture nor an over-pessimistic and cynical picture. That spokesmen for corporations will talk a lot about the public interest but actually act to protect their special interests, can simply be taken for granted.

If you don't have a realistic picture of the action situation on a major public health issue, then any action you try to take is likely to do more harm than good. The action may have the best of intentions. However, good intentions are no substitute for a clear understanding of what is going on in these complex technological issues. For example, when I first started out in the smoking-and-lung-cancer issue more than 25 years ago, I firmly believed that the protection of the public was a job for health professionals. I felt that the ordinary private citizen should not have to worry about the technical matters involved in health and health care--that it was up to the professionals to do the worrying and to solve the problems. Indeed, I shared the widespread prejudice of my colleagues that the public would do just as well to keep its nose out of affairs that it couldn't understand and could only further confuse. Although my good intentions were tinged with the elitism that permeates professional communities, I meant well. But I was wrong and it was a very costly mistake.

From 25 years of experience in one public health issue after another, I have learned that the public cannot afford to rely on professionals to protect it from hazardous technological products and processes. A substantial and influential segment of each of the professional communities--of the scientists, doctors, university professors, health agency staff, and other professionals--are more concerned with the protection of technologies than with protection of the public. For example, in 1977 I wrote an article which advocated that with potentially hazardous technologies, the benefit of the doubt should go to the public and not the technology, and submitted it to *Environmental Health Perspectives*. Every single reviewer indignantly rejected this notion (although the article was eventually published as a letter).

It is natural for persons who are themselves involved with some area of science or technology to be more sympathetic with fellow professionals in other technological areas than with the public--it is one of the folkways of all professional groups that "insiders" stand together against "outsiders." This "fact of life" makes it easy for the promoters of hazardous technologies to get allies and apologists from various professional communities. It makes it hard for the public to get defenders from these communities.

Therefore, it should not be surprising that over the years that I have been fighting for effective measures to protect the public health, my adversaries have generally been professionals rather than corporate executives or nontechnical staff. In other words, the persons who have done the most to block or delay needed action to protect the public have generally been scientists, university professors, doctors, administrators of prestigious institutions, health bureaucrats, and other professionals.

This is not a blanket indictment of all professionals. The "silent majority" of practitioners have stayed out of these controversial issues completely. A substantial number of professionals have spoken out courageously to protect the public (despite peer pressures to remain silent). I have noticed that professionals who deal with the public on a one-on-one basis are likely to feel some personal responsibility for protecting the public.

It is frequently the professionals in the upper echelons of their communities (or ambitious younger persons who are trying to get into these echelons) who show no qualms about acting on behalf of special interests and against the public interest. It is professionals whose business is trading in administrative "brownie points" who show a total dedication to their own self-interests (or the interests of those who can return the brownie points at a later date) and a total indifference to the interests of all other human beings. This is a grim fact-of-life to remember about the top brass of the professional communities, particularly since these same people are also the most eloquent speakers. Whenever they want the public's money, they are not shy about posing as white knights in shining armor.

STEP 2: GETTING INVOLVED

My main purpose in introducing some of the leading actors in the public spectacles on health issues that have been going on for the past 25 years is to explain why I changed my mind about "letting the health professionals do the job." For one thing, when we consider the overall protection of the public health and safety against hazardous uses of technology, we would have to say that the professionals have generally failed to do their job. Indeed, about the only time progress has been made in stopping the abuses of technology was when "concerned citizens" made a substantial effort to get something done.

One of the lessons of 25 years of experience, then, is that the ancient rule still applies: The gods help those who help themselves.

To put it another way: If you want to save your own life (or if you want to protect the health and safety of those who are near and dear to you), then you will have to take a more active role. In these complex issues, action for its own sake (or for the sake of public relations) is likely to do more harm than good. To act effectively in your own interest, you have to have some understanding of the action situation for a given health issue. Like it or not, you will have to learn about some technical matters. For this purpose you need to know something about how a working scientist tackles the job of finding out what is producing a health hazard and what can be done to control it. You need to know some general strategic guidelines and some specific information. This book is designed to help you help yourself.

Perhaps you are not a do-it-yourself enthusiast--I'm certainly not. Whenever something goes wrong around the house, my reaction is to call a professional in to fix it. For anyone as unhandy as I am, it is usually cheaper and better to have a professional make a minor repair immediately than to do the major repairs needed after I've botched the job. A major public health issue is a far more complicated matter than fixing a faucet, and so it is understandable that many readers would be somewhat reluctant to get involved in such matters. Hence, it is not surprising that in the past generation the American public and the citizens of other technological societies have put their faith in technologists. It has seemed self-evident (an impression which has been encouraged by the technologists) that the decision-making on how technologies should be used should be controlled by those who have the technical expertise.

Logical as it may sound, it has been a disastrous mistake to let technologists make the decisions on how technology should be used (or, as it is now fashionable to say in technical-sounding jargons, how it should be "deployed"). It is important to distinguish between the pushing of the buttons or the setting up and maintenance of equipment (which is the *operation* of the technology) from the benefits or costs to the public that result from these operations (which is what matters in the *use* of the technology). Technologists have expertise in the *operations*, but they are not experts on (and know surprisingly little about) the problem of effective deployment. Many take the simplistic and very dangerous view that "the more technology the better"--a view that has benefits to the technologists but has often had fatal consequences for the public.

It might be noted in passing that there is now coming into existence a technology which can be used to facilitate the safe, effective, and economical deployment of technology. It can be called "metatechnology." I will say more about metatechnology later. At this point, I will merely note that very few technologists have any competence in this area and that the skills required are completely different from those needed in the operation of technology. So you see that it is not as logical as it sounds to have technologists make the decisions about the uses of technology. Nor is it so surprising that decisions made by technologists have often been so bad for the public health and safety.

A basic fact of life about the sophisticated and complex new technologies is that no one really knows how they are going to work or how they are going to fail. About the only prediction that can safely be made is that there will be some unexpected, and probably undesirable, consequences or side effects--just what they will be cannot be predicted. When these adverse effects start showing up, they are likely to be dealt with if there is a "technological fix" that will take care of them. If not, there will be pressures to sweep them under the rug so that they cannot be exploited by rival technologists, environmentalists, or commercial competitors. When technologists make the decisions about the uses of their technologies, they will protect their own interests (as they see them) rather than the public interest. It is a "fact of life" that professionals are no more altruistic than anyone else.

At the lower levels there have been technologists who are courageous enough to blow the whistle on hazards to the public, even at the cost of their careers. However, in the upper echelons, where the key decisions on the deployment of technology are made, altruism is regarded as a crime. For the past 25 years, the public has put its trust in professionals--particularly the top brass in government agencies, corporations, educational and research institutions, professional societies, and voluntary health agencies. The sad truth is that the upper-echelon professionals have grossly abused or betrayed this public trust.

At this point, you might be wondering: If technologists don't make the decisions about the deployment of technology, who can do this job? My answer would be: You can. Indeed, if you are going to save your life, you will virtually be forced to become involved in this decision-making.

This isn't as difficult or impossible as it might seem at first. In recent years, a fairly large number of "concerned citizens," working alone or in small moneyless groups across this country, have become surprisingly expert on extremely complicated technical issues such as the nuclear power controversy. At recent meetings that I've attended, the questions asked by persons with no formal scientific background concerning specific federal regulations have been so very technical that none of the experts from the regulatory agencies knew the answers. At a recent Nuclear Regulatory Commission (NRC) hearing (that I'll be saying more about later), a young lady from an environmentalist group took on a Harvard professor on his own terrain. Despite his put-downs and evasive answers to her well-organized series of questions, she backed him into a corner and forced him to admit that what he had said was wrong.

Not everyone can out-expert a professional. However, anyone can learn to recognize the linguistic tricks that the experts use to "put the public in its place" and can refuse to be intimidated by these tricks. Incredible as it may seem to some readers, you will not be struck by lightning for blasphemy if you talk back to a professional. Indeed, another fact of life is that the public cannot get good service from professionals by being "properly" deferential--the only way to get good service is to take no nonsense from the experts.

STEP 3: KNOWING WHERE TO START

Perhaps the hardest part about getting involved in the complexities of a major public health issue is knowing where to start. There are so many different aspects of any real issue and they are so interrelated that one is always caught in a Catch-22. In order to understand any one particular aspect, it is first necessary to understand half a dozen others.

There is really only one way out of this dilemma: Jump into the middle of all the complexities and hope that with some hard work, all of the confusing aspects will begin to come together and make sense. It is rather like learning to swim by jumping into the middle of the pool. If you can keep barely afloat for a while, you will eventually be able to navigate.

Anyone facing these complicated health issues for the first time may envy the experts who seem to move so easily through all of the difficulties. However, the ability of experts to cope with all of these complexities is largely an illusion that professionals go to some pains to create. *The truth is that no one can ever be fully expert on any of these major issues*. They have entirely too many facets for anyone to know them all. In my experience, there have been very few alleged experts who have any broad grasp on the issues. However impressive they sound, few have any real competence--at least in those areas where I am a pro and can judge performance.

As a professional statistician, I have listened to a great many experts quoting statistics. They are glib and persuasive so far as an audience of nonstatisticians is concerned. Actually, these experts know as little about statistics as anyone in the audience. A certain air of assurance, or even arrogance, is all that is needed to carry off this deception of the public.

The professional often has only one real advantage over a concerned citizen in the discussion of a public health issue. This advantage is: Language. Whereas the public speaks ordinary English, the professional can speak one or more of the technical sublanguages of English. The ability to speak a technical jargon provides the expert with much more than ready access to a body of information that is otherwise not very intelligible. The jargon is at once a means of identification, a membership card in a restricted club, and a sword and a shield in any confrontation with other professionals or the public. It can be an overwhelming linguistic advantage if an adversary is unfamiliar with the gimmicks that are built into technical jargons. It is easy to make you feel helpless or discomforted even when you really know more about the matter than the so-called "expert."

On the other hand, once you recognize that this is just a language game, that rattling off technical-sounding statements is no proof of actual competence in these public health issues, then you don't have to be overawed or brow beaten. Do as the young environmentalist did at the NRC hearing--keep boring in with questions in plain English. If you persist you can watch the icy professional image start to melt away in the anger and confusion of an "expert" whose expertise is publicly questioned. Of course this won't work with a real expert because such a person can give sound answers in plain English. However, in these public health issues there is not much danger that you will have to confront a real expert.

The technical jargon of a professional serves many purposes, some technical, but many sociological in nature. One way it helps in approaching a major issue is that it excludes from consideration all aspects of that issue that are beyond the scope of the jargon (and technical jargons have a very narrow scope). The professional, therefore, has much less difficulty with complex public health issues because the jargon of a specialty sets the limits of where to start and where to stop. A professional begins and ends with those aspects of the overall health problem which can be talked about in a particular jargon. From the narrow viewpoint of a technical specialty, anything outside of its boundaries cannot really be of much consequence.

For anyone who does not wear the blinders imposed by a technical jargon, the many different aspects of a public health problem can be a psychological barrier to getting involved. There are dozens of questions that a cautious person venturing into a major health issue might want to have answered before getting in too far. These range from very broad, strategic questions to fine points of technical detail. For instance: What are the strategic priorities in a health issue? What are the ground rules for judging the validity of scientific arguments? What are the general areas of agreement or disagreement concerning the hazards?

At the other extreme are specific details of a particular health issue. For the hazards of low levels of ionizing radiation that will be used as a continuing example here, there are immediate questions about technical language: What is a rad? What is a rem? One can give a technical answer that will merely bog you down in further technicalities (this is a favorite linguistic gimmick for locking the public out of these issues). But there is also a nontechnical answer: Rads and rems are measures of amounts of radiation. For the kind of radiation that will be considered here, there isn't much practical difference between the two.

What is "low-level" ionizing radiation? Again the answer can be informative and get to the facts, or it can be "correct" in a technical sense but merely obscure the issues. For instance, "low-level" radiation can be taken here as between 100 millirads and 10 rads (two points on the yardstick used to measure the amount of radiation absorbed) or what can be called, for brevity, the "1-rad range." Why is this particular range chosen? Simply because the vast majority of all of the exposures for nuclear workers or for ordinary diagnostic radiation occur in this 1-rad range. This is the crux of the public health problem.

To muddy the waters, this kind of answer would be given: "There are differences of opinion about what constitutes 'low-level' radiation, and the designation is arbitrary." This can be the jumping-off point for a long harangue that will actually say nothing consequential. In these issues, one can almost always say that there are "differences of opinion" or that terms will be somewhat arbitrary. If one wants to talk sense instead of rattling off jargon, however, there are practical reasons for distinctions such as "low level" and "high level." These will become clearer as one gets further into the issue.

So it is important not to be put off at the start by the pompous nonsense that is used by professionals to "put the public in its place." There are many questions that are troublesome at the start, but tentative answers will carry you along to the point where more comprehensive answers can be given and can make sense. Getting "into" a major health issue is an evolutionary process where you learn as you go and where it can take years before you have a sure grasp of the problem. This is how I have to do the job even after more than 25 years of experience in this area. Those so-called "scientists" who sound off on these health issues without bothering to learn about them are simply faking it. One of the major obstacles to saving your life is that most of what is said by persons who speak "in the name of science" is an out-and-out deception--often self-deception. In less polite terms, professionals have been lying to the public for years, and if you want the truth you will have to look into these issues for yourself.

The next chapter will test out the theory that the best way to get started is to jump into the middle of a real issue. In this and subsequent chapters, the main line of the text will follow my presentation on April 7, 1978 to the NRC of our major new findings on the hazards of low-level ionizing radiation.

This all-day hearing for this specific purpose followed the general lines of a format that is called "Adversary Science." The scientific data, methods, and results were presented in plain English and in common sense terms which are understandable to the general public. Then there was a prepared critique of this work and an informal rebuttal of the critique.

In an Adversary Science format there would then be cross-examination. The NRC hearing didn't go that far. However, the cross-examination actually occurred when, in effect, there was a three-hour question period from the audience. I was cross-examined by the members of the "radiation protection community" who were there in force (but weren't too troublesome because they hadn't bothered to do their homework). They, in turn, were cross-examined by the "concerned citizens" of the environmentalist groups (who had done their homework), and they had their troubles.

chapter 2

SCIENCE IS COMMON SENSE REDUCED TO CALCULATION

The basic scientific strategies in public health: The Galilean rule and the Primacy Principle.

INTRODUCTION

This chapter and Chapters 4, 6, and 7 contain the material that I presented to the Nuclear Regulatory Commission (NRC) in an attempt to get this federal agency to take action to protect the workers and the public from the health hazards of low-level ionizing radiation. The dosage of radiation which had been permitted for many years by the NRC (and its predecessors) is capable of doing serious harm, and my job was to try to further reduce the permissible level by a factor of 10 to a less hazardous level. My role, as I saw it, was *not* to present my opinions. Rather it was to act as a spokesman for the mass of factual evidence that I was presenting. These facts show a definite, serious human health hazard at the dosage permitted, and the regulations actually allow this dosage to be given to the worker *every year* that he works at an installation (though in practice this doesn't often happen).

Given this strong, scientific evidence from a population base of 13 million humans and from hundreds of cases of leukemia and other diseases, one might expect a federal agency whose reason for existence is to protect the human beings to take immediate action. We will see later what actually happened. It might be enough to note that if federal agencies did their job, if technology-oriented corporations accepted their responsibilities, and if the professionals involved with technology behaved like human beings instead of like technocrats, then it would not have been necessary for me to write this book. And now for the actual presentation that was made on April 7, 1978 in Washington, D.C.

BROSS TESTIMONY BEFORE THE NUCLEAR REGULATORY COMMISSION (1978)

STRATEGIC GUIDELINES: THE PRIMACY PRINCIPLE

My intention at this hearing is to present the complex technical issues involved in the study of the hazards of low-level radiation in language that the general public can understand. We are using a format sometimes called "Adversary Science." Because this is one of the first hearings of this kind on a technological issue, we have few guidelines or ground rules or precedents, so I will start by suggesting one.

This guiding principle might be called the "Primacy Principle": When we are dealing with a public health issue involving possibly hazardous technology, the benefit of the doubt must go to the public and not to the technology. In the past, regulatory agencies often operated on the reverse principle, and this is why we now have so many environmental health problems.

The Primacy Principle allows us to deal with more specific issues such as the question of "burden of proof." In these complex issues it is unrealistic to expect that the hazards of a technology can be proved beyond a shadow of a doubt. There always will be some doubts. Applying the Primacy Principle we are led to the following practical corollary: The proponent of a hazard hypothesis has the burden of establishing a prima facie case for the hazard. Thereafter, the burden of proof shifts to those who want to use a technology to show it is safe.

What I propose to do here is to establish a strong prima facie case that there are serious human health hazards from dosages of ionizing radiation in the range between 100 millirads and 10 rads. For brevity, I'll hereafter refer to this as the "1-rad range." The current standards set by the Nuclear Regulatory Commission permit exposures up to 5 rems per year, that is, exposures in the 1-rad range. I will argue that nuclear workers should not be exposed to known hazards, and the 5-rem level should be reduced by a factor of at least 10.

To make my case I will present factual evidence from a population base of 13 million persons. I will be dealing with matters of fact, not matters of opinion. It is a matter of fact that persons reporting exposure to trunk x-rays at these low levels of ionizing radiation showed increased risks of leukemia and other diseases. In some of the data the effect shows up in the person exposed, in other data the effect shows up in the child of the person exposed. These facts establish a strong prima facie case that there are serious human health hazards in the 1-rad range.

HISTORICAL CONTEXT: SCIENCEGATE

In view of recent events, I will depart from my previously prepared text to try to put what I have to say into a broader historical context. In January and February of this year, the House Subcommittee on Health and the Environment held a series of meetings on the actions of "Big Science" federal agencies that have covered up the health hazards of low-level ionizing radiation for the past 20 or 25 years. One example of what might be called the "Sciencegate cover-up" was the testimony on February 8 and 9. This deals with the efforts of the Department of Energy [DOE] to suppress possible positive findings on radiation effects on Hanford workers in the study of Dr. Thomas Mancuso--whose work I'll mention later. Chairman Paul Rogers called this "the worst mess" he had ever seen in a research area. Under agencies such as the Atomic Energy Commission [AEC],

Energy Resource and Development Agency [ERDA], Department of Energy, and Nuclear Regulatory Commission, the Sciencegate cover-up suppressed positive findings, tried to discredit their authors, and finally cut off their funds to stop the research. The details of this dirty business are in the Congressional Record.

The NRC is a hardware-oriented, not a health-oriented, agency. This hearing is being held today despite the NRC's efforts, not because of them. Two years ago NRC promised this meeting and then reneged. A year later, after pressure from the White House and Congress, NRC finally agreed to an immediate hearing. That was a year ago. Perhaps we can dispense with the polite fiction that NRC is neutral on this health issue or at this hearing.

Let me give just one example of the one-sided NRC arrangements. Last year, after I had sent in my scientific report in the form of slides, Dr. Kenneth Rothman asked for my written text. I didn't want to send this in, but NRC refused to set a date for this hearing until I did so. After a couple of weeks, I capitulated to avoid further delay.

Recently I asked to see Dr. Rothman's written text, but this was refused. In his letter of March 8, 1978, Dr. Rothman gave this explanation to the NRC:

"I trust that you and Dr. Bross appreciate that these ten weeks of preparation are necessary for me to complete a satisfactory review of the material. I have had only a slight familiarity with radiation carcinogenesis prior to this assignment, whereas Dr. Bross has been immersed in this field, and thereby preparing himself on these issues for many years."

I will have to concede that Dr. Rothman has a good point here. But his real handicap is not my decade of research on x-ray hazards or even my 25 years as a professional biostatistician and public health scientist. My real advantage is that my case is entirely based on solid factual evidence--on data on human beings actually exposed to low-level radiation. This evidence is so strong that even if a fairly competent analysis is made, the radiation effects will come through. When a good analysis is made, the effects shine through so clearly that no one but a Sciencegater could deny them.

SCIENTIFIC STRATEGY: ASSESSMENT OF HAZARD

So let me now go directly to the scientific question: "Can exposure to low levels of ionizing radiation in the 1-rad range produce serious health effects?" At a recent congressional seminar on this topic where a wide spectrum of views were represented, every competent scientist present said "yes." A few representatives of the radiation protection community retreated into their usual double-talk, but no one stood up and said "no." This was February 10, 1978. About two years ago at a similar seminar it was a different story. The Sciencegate cover-up began to collapse about that time, and today the public and its representatives are well aware of low-level radiation hazards, even if some regulatory agencies seem reluctant to admit the truth.

The real question today, then, is "How much of a hazard is there?" This is a quantitative question and today I will give a quantitative answer. To get to this answer without delay, I will reverse the order of my presentation and present first the facts about nonlymphatic leukemia in men exposed to radiation in the 1-rad range.

Since I want to present our research findings in a way which will make sense to the general public, let me begin with the question: How does a scientist get numerical estimates of the health hazards of low-level ionizing radiation? Although the task is difficult, the strategy is simple and straightforward. For example, one way to do this job is to produce what is called a "dosage response curve." My first slide [Fig. 1, Slide 1]* shows what a dosage response curve looks

*The figures begin on page 87 at the end of Chapter 7.

like and in fact shows three different theoretical curves (marked A, B, and C, respectively). Notice that on the horizontal axis the dosage levels are marked. In the graphs that I will be showing, the dosage will be given in rads, a measure of radiation that has already been introduced, and the range of dosage shown will generally be the 1-rad range or a little higher.

The vertical axis shows the "response variable" which is some measure of the biological effect of a given dosage. I will use as response variable an estimate of what will be called the "proportion affected." Later on, I will explain how we get this estimate. At this point I will merely note that this can be regarded as the proportion of the persons exposed to the x-ray who suffer some permanent genetic damage that may be expressed as leukemia or other disease. If low-level ionizing radiation is producing such damage, then we would expect that the proportion of persons affected would increase as the dosage increases. The slide shows that this might happen in different ways, and much of the controversy over low-level radiation hazards involves the question: Which is the correct curve, A or B or C?

ANATOMY OF A CONTROVERSY: THREE THEORIES

The controversy has continued for years because no one was able to produce an actual dosage response curve for the 1-rad range. All of the dosage response curves were for data in the centirad or kilorad range--for exposures around 100 or 1000 rads. Today I'll show you the first curve for the 1-rad range. It ought to settle the controversy but probably won't.

Once a dosage response curve is available, it is not difficult to provide quantitative measures of hazard. One is called the "doubling dose" and is related to the slope of the dosage response curve. We have estimated the doubling dose and it turns out that it is probably less than the 5-rem dose which the NRC permits annually. So the hard part of the job was to get good estimates of the proportion of those exposed who are affected by radiation in the 1-rad range, and once this is done the quantitation of the hazard is largely a computational chore.

While we are looking at this theoretical dosage response curve, let me tell you what the controversy was all about. The crux of the matter here is the use of the procedure called "linear extrapolation" to estimate the hazards in the 1-rad range from data that were mostly in the centirad range. Extrapolation outside of the range of the actual data is always a dangerous thing to do in statistics or in science. It was doubly dangerous here because the extrapolative estimates in the famous Report of the Advisory Committee on the Biological Effects of Ionizing Radiation [BEIR Report] have been widely used in the setting of standards. As we'll see, they seriously underestimate the hazard. They are not 10 percent or 20 percent, but more like 1000 percent off.

For all the technical-sounding double-talk that has been used to "justify" linear extrapolation from high radiation doses to low doses, the procedure is nothing more than an educated guess that depends on a very strong assumption--the assumption of linearity. The controversy involves the challenge and defense of this assumption. It is, however, somewhat misleading to put all the emphasis on whether or not the curve is a straight line because it is not simply departure from linearity that matters. We must ask ourselves: What does such a departure imply? Some departure from linearity can lead to a serious overestimation of hazards, others to serious underestimation, and still others to a relatively minor bias in the estimates. What is really crucial to the validity of the extrapolation is the way in which the dosage response curve behaves when the dose gets down into the 1-rad range.

This hypothetical set of dosage response curves illustrates this point. If the dosage response curve looks like the one shown in A [See Fig. 1, Slide 1], it turns sharply downward as we get to the 1-rad level. This would imply that there was a threshold or cutoff, a claim which has often been made but for which there is no good supporting data at all. If the curve looks like B, which you see continues as a straight line, then of course the linear extrapolation will give a correct estimate. Finally, if the curve flattens off as in C, as it gets down to the 1-rad range, this is from the public health standpoint the worst possible case. If this happens then the extrapolative estimates can seriously underestimate the actual risks in the 1-rad range. We will be able to settle this argument when we see the actual curves for the dosage response. You've just seen the theories. When I show you the facts it will be easy to see which theory is probably right.

SCIENTIFIC STRATEGY: RESOLVING CONTROVERSY

Now that you've seen the theory, the next slide [Fig. 2, Slide 11] shows you the facts. The horizontal or x-axis shows the dosages of radiation delivered. The vertical or y-axis shows the "proportions affected" by the dosage. The particular series shown here are men over 65 in the Tri-State Survey. I'll tell you more about this data, which is probably the best of its kind, a little later. We are, however, talking about men, not mice, and we are talking about men exposed to dosages in the 1-rad range and not dosages of hundreds of rads.

The graph not only shows the estimate of the "proportion affected;" it also shows the precision of this estimate. This is shown by the 95 percent confidence interval on the proportion affected. The intervals show how much the estimates might be expected to vary in repeated samples. We can see, for instance, that at the lower doses about 5 percent of the individuals are affected by the x-ray. This is probably not a random group of persons, but rather individuals who are particularly susceptible to the x-ray because of preexisting genetic damage. I would speculate that they have more breakpoints in their DNA than most other persons. If the x-ray produces one more break, these people will be affected.

As the dosage increases, the proportion affected tends to increase. The shape of the curve here is *not* something which is forced into the data by any assumption such as a linear assumption. The curve can go anywhere it wants to. As you will see, it occasionally goes down slightly, but usually goes up. When you look at the confidence intervals, you can see quite clearly that it tends to go up. You can see the lower limits of some of the confidence intervals are above the upper limits of other confidence intervals. There is no question that there is a positive slope, but the curve of the slope seems to be changing rather than constant.

Previously I said that the critical question that we would try to answer would be: What happens to the dosage response curve as you get down into the 1-rad range? Does the curve suddenly drop off, does it go down very sharply? Does it flatten off? Does it stay linear? Those were the theories. Now you have seen the facts. The confidence intervals don't entirely distinguish between a straight line and other increasing curves, but we can certainly rule out the possibility that the curve drops down to zero at any point. So the notion that there is a threshold is flatly ruled out.

This is very important. The notion that below some so-called "threshold dose" radiation becomes harmless is clearly contrary to the facts. The technologists who have often made this claim have jeopardized the lives of hundreds of thousands of Americans and the health of their children by insisting on this self-serving myth. The myth justified the promiscuous use of radiation at the dosages at the 1-rad range. Those who asserted this myth "in the name of science" were lying to the American public.

The next slide [Fig. 3, Slide 13] is a bit more complicated. This is because it shows all of the age groups instead of just the oldest men. Since these estimates are all independent, they serve to confirm the previous slide. The dosage response curve is rather similar in all of the age groups. From this we can draw some conclusions now.

First of all, the threshold hypothesis is flatly contradicted by the facts. There never was factual support for this hypothesis and there is no excuse any longer for anyone in the radiation protection community, particularly anyone involved with standards, foisting this exploded notion on the public. What about linearity? This is a little tricky. We can draw straight lines which go through all of the confidence intervals in the 1-rad range. Hence, strictly speaking, we cannot rule out the possibility that the curve is approximately linear in this range. When we get above the 1-rad range it is much harder to put a straight line through the confidence intervals, and it seems fairly evident that the linearity will not continue all the way out to hundreds of rads. We therefore should not try to extrapolate these results too far up the dosage scale. The reverse is also true--we shouldn't try to extrapolate down from the high-dose range.

THE PRIME DIRECTIVE: THE GALILEAN RULE

The actual dosage response curves for the 1-rad range show why the BEIR estimates have missed so badly. The curve levels off in the lower ranges, probably because we are getting into a susceptible subgroup of the general population. It does not go down to zero as it should with linear extrapolation. For persons who are more familiar with dosage response curves where there is a nonzero baseline level, let me stress that the analysis used here has already taken into account the baseline level of leukemia. Therefore, this graph should theoretically go through zero in the complete absence of radiation.

The dosage response curves show why the BEIR estimates are off. To deal with the point more directly, we can use the "doubling dose" that is often used in quantitating radiation hazards. I personally have reservations about the use of this quantity. However, it is widely accepted and it does help to make my point in a more quantitative way. Mathematically, there is a simple relationship between the doubling dose and the "percent affected" that has been used in the dosage response curves. Since there is a mathematical model which I'll describe a little later, it was possible to set up a computerized procedure for estimating the doubling dose. The technical details are not very interesting or important, but I should say a little about the scientific strategy that is used here.

The prime directive of modern science is the Galilean rule: A theory must fit the facts. The Galilean rule has been the hallmark of genuine science ever since Galileo first used it to demolish the Aristotelian theories of motion. Strict adherence to this rule is the mark of a real scientist, and anyone who deviates from this rule is not functioning as a scientist no matter how many Ph.D.'s or M.D.'s he may have.

The Galilean rule is the scientific counterpart of "honesty is the best policy" and is sometimes a hard rule to live by. It can also be a difficult rule to apply, and this is where quantitative methods often help. If the facts can be expressed as numbers and if a theory gives numerical predictions, then seeing whether a theory fits the facts can sometimes be reduced to a numerical operation.

STATISTICAL TACTICS: USING THE GALILEAN RULE

The way this works in practice is, for instance, that we get a count of the number of persons in a particular category from the Tri-State Survey. This can be a very complicated category such as, for example, "the number of men over 65 years who were exposed to radiation less than 1 rad, who had heart disease, and who also had a diagnosis of leukemia." The actual count for this category is nine. This count is often called an "observation."

We also have a theory which is expressed algebraically in a mathematical model and arithmetically in a computer program.

From this theory we are able to predict what the count will be in a given category. This prediction is often called an "expectation." This number might be 8.27. If we want to see whether the theory fits the facts we can look at 9 and 8.27 and notice that they are similar. If we want to be more quantitative we can calculate the numerical difference between observation and expectation,

$$9 - 8.27 = 0.73$$

and use this in an index of goodness-of-fit.

The technique is called a "chi-square goodness-of-fit test" and it involves some additional arithmetic. The difference is squared, divided by the expectation, and summed over the categories to obtain a number called "total chi-square." None of these technical details are very important. The main point is that we can go through the data tables that I'll show you shortly and obtain a quantitative measure of just how well a given theory fits the facts.

The next slide [Fig. 4, Slide 14] shows how we can use this strategic approach to get a numerical estimate of the doubling dose. The x-axis shows a range of possible values for the doubling dose. The y-axis shows the corresponding values of chi-square. A lot of calculation is involved, but this is easy to do on a computer. What you notice is that the curve starts out high, decreases until it reaches a minimum, and then goes up again. Remember, we are using a quantitative measure of goodness-of-fit between theory--here a particular value of the doubling dose--and fact--here a table of actual data. Where the minimum occurs, this is the theoretical doubling dose that best fits the facts. It is plausible to use this value as our estimate of the doubling dose.

The horizontal lines in this slide are used to get approximate confidence intervals on the estimate. Where the curve intersects these lines, the goodness-of-fit has become so poor that we would say that particular value of the doubling dose is unlikely.

Sometimes a single observation may be so out of line that it distorts the estimate. To check this point, the analysis was redone dropping one of the 2 × 2 tables. The effect was to reduce the estimate of the doubling dose.

All in all, this slide [Fig. 4, Slide 14] tells us that the doubling dose is probably less than 5 rems. It also tells us that it is extremely unlikely that the BEIR Report estimates of doubling dose, which range from 25 to 50 to 100 rads, are correct. Quantitatively it tells us that the BEIR estimates are off by an order of magnitude. Hence the use of these estimates will not protect the public or the workers, and this is why we must cut the 5-rem level by at least a factor of 10.

One reason for estimating the doubling dose was to get a direct comparison between the work of Mancuso, Stewart, and Kneale that was based on the exposure of Hanford workers to nuclear radiation and our study. Their first estimate of the doubling dose was less than 1 rad, which was incompatible with ours. However, the estimate that Kneale has just reported in Vienna is 3.6 rems, which is in good agreement with ours. These mutually confirmatory results from two completely independent studies that used different data, analysis, and approaches is of some scientific importance-particularly since they are the *only* two studies which deal with massive series of human beings actually exposed to radiation in the 1-rad range. It is also important that the different kinds of radiation--in one case nuclear radiation to workers and in the other diagnostic x-ray to the general public--seem to have similar health hazards. The two studies are complimentary in the sense that each fills in a gap in the other study. In our case the Mancuso findings show that the x-ray findings carry over to the nuclear radiation of primary concern to the NRC. At the same time our much larger number of leukemias reinforces the Mancuso findings that involve fewer cases.

Our quantitative results on radiation-induced leukemia in men are much stronger than the prima facie case originally promised. Taken together with the Mancuso findings on Hanford workers and the findings on childhood leukemia which will be presented, the results are close to definitive. There is certainly more than enough evidence here for federal agencies to take immediate action to protect workers and the public against unnecessary exposures in the 1-rad range.*

---

*Further information on the studies by Dr. Mancuso mentioned in the NRC presentation and on other research on the effects of low-level ionizing radiation on humans actually exposed to such radiation can be found in the second section of Chapter 11, and in the references given at the end of Chapter 11.

chapter 3

GIVING YOU TOP PRIORITY: THE PRIMACY PRINCIPLE

Public health action in the face of doubt and uncertainty: Giving humans, not technology, the benefit of the doubt.

STRATEGIC PUBLIC HEALTH GUIDELINES

Let me pause at this point in my NRC testimony to take a closer look at one aspect of this testimony that has an immediate bearing on the question: What can you do to save your life from technological hazards that you cannot (as one individual in more than two hundred million) control?

As one individual you may feel helpless, but there are millions of other persons who are becoming very concerned about the inadequate safeguards to their health and safety. If you act together with others, you can demand (and I believe get) a congressional resolution directing the federal agencies to give top priority to protecting your life in the regulatory actions taken by these agencies. A resolution affirming the Primacy Principle that was introduced in the last chapter would not, of course, solve all of the problems. However, it would provide some encouragement in the federal agencies for those bureaucrats who are on your side (unfortunately a minority) and some leverage against those bureaucrats who are not. Furthermore, it would put your representatives in the Congress on record as to whether they are more interested in protecting your life or protecting their campaign contributions from corporations that jeopardize your life with unsafe technology. That is something worth knowing.

What is the Primacy Principle? Why is it needed? Why has it (and some guidelines related to it such as the Delaney Amendment) been so bitterly opposed by so many professionals? Why are the upper-echelon doctors, scientists, and other professionals, who ought to be on your side, lined up on the side of those who are jeopardizing your health and safety? In December 1977, after a bit of a hassle, *Environmental Health Perspectives (EHP)* published a letter which gives some answers to these questions.

Some of the letter will be a little repetitive. However, I believe that it will be useful to some readers to see the original working documents that were used to try to get public health action to save your life. They are not beautiful, but they illustrate how technical subjects can be discussed in plain English.

I might note that you as a taxpayer are paying for journals such as *EHP* (which is a government publication). However, the "referees" who decide what goes into these journals are not greatly concerned about the public interest and often have ties to the technologies that are creating the public health hazards. It is a system which makes it difficult to publish the information that could save your life.

LETTER TO THE EDITOR:
THE PRIMACY PRINCIPLE

Dear Sir:

Recently I submitted a paper to EHP proposing a guideline for action to protect the public against environmental hazards arising from technological products and processes. I called it the "Primacy Principle." To implement this guideline, I suggested a special format and review procedure for papers presenting new data on technological hazards. The EHP reviewers didn't care for my proposals and unanimously rejected my paper. As one reviewer remarked, "I doubt if EHP, as a purely scientific journal, should go into this administrative and political field."

There was some suggestion that the paper might be more appropriate if it were recast as a "Letter to the Editor" or presented at an American Public Health Association conference where a "more immediate consensus" on the proposals would be possible. In what follows, I have tried to follow both suggestions by rewriting the material as a letter and incorporating some of the editorial reactions. Therefore, let me start with the original statement of the "Primacy Principle":

With potential technological hazards to the public health, the "benefit of the doubt" must go to the public and not to the technology.

There seem to be two main objections to the Primacy Principle. One criticism is that it is not "scientific," that it is "political" or "subjective." I agree. When (as here) we are dealing with the setting of priorities, we are necessarily outside of the boundaries of "pure science." We are dealing with the value systems of human beings. However, I would argue that environmental health is not "pure science." There is no good reason for the public to support such research unless the findings are translated into action programs to protect the public from environmental hazards. Effective action involves setting of priorities. Therefore, the topic, "scientific" or not, seems to me appropriate for a journal called *Environmental Health Perspectives*.

Another objection involves the concept that science is "neutral," that it is neither on the side of the proponents of a potentially hazardous technology nor the side of the public. Insofar as the estimates of hazard are based on sound statistical methods applied to reliable factual evidence--insofar as the estimation is scientific--then the estimates can be objective, unbiased, and neutral. However, it is a fact of life that in any complex actual health issue we will not have scientifically valid estimates of all the relevant benefits and costs. To take action, we eventually must do some guessing. When we talk about the benefit of the doubt, we are talking about guesswork, not science.

Guesswork is inherently subjective, and this is true even when it is called expert opinion or linear extrapolation or risk-benefit analysis. This is where the benefit of the doubt comes in. When we guess, we can make mistakes in either direction. The Primacy Principle says we should prefer to fail in a direction which will be fail-safe for the public. When we have to guess, we explicitly give the public priority over technology.

A different kind of objection to the Primacy Principle is that it is too vague. This vagueness is inherent in general guidelines and can be reduced by exploring the implications of a general principle. By applying the principle to the troublesome questions about the "burden of proof" in publications on potentially hazardous uses of technology, for example, I suggested the following amplification: Under this principle, the burden of proof is on the health scientist to produce prima facie evidence of hazard. Once this is done, the burden of proof shifts to the proponents of a technology to show it is safe.

The evidence needed for a prima facie case is analogous to that required for a grand jury indictment (as opposed to conviction at an actual trial). Thus it is sufficient to present an outline of the data, the methodology, and the conclusions to show that there is a case for the existence of the hazard. The case itself (with full details and discussion) would presumably require a hearing and could not be presented in a single paper. Where possible, references to evidence should be cited rather than reproducing all the material in detail. The argument or rationale should be in plain English rather than in a technical or scientific jargon.

LETTER TO THE EDITOR: APPLYING THE PRINCIPLE

To implement this working rule for determining the burden of proof, some expeditious way of communicating the early warning report (and the prima facie case) to the health scientists and to the public is needed. What I suggested was that some journal (not necessarily *EHP*) regularly print such early warning reports. This does not mean that any and every warning would be printed. There would have to be new data (or new analyses of previous data) presented that might be considered a prima facie case for environmental hazard. Neither the journal nor the referees would have to accept or endorse the findings. These points could be made very clear in explanatory text introducing the Early Warning Report. It would be up to the readership, and ultimately the public, to judge whether or not a prima facie case had indeed been made.

Since the suggested format for an early warning paper is somewhat different from the traditional formats (for instance, it would not include any extensive literature review), there would have to be some special editorial processing. Rejection should be based on something specifically wrong with the data or the methodology (and not on general objections or doubts). Precautions would have to be taken against having proponents of the potentially hazardous technology (or their scientific allies) block the publication of early warning reports.

The reaction of the referees to these suggestions for implementing the burden of proof proposal was extremely negative, and I have had similar reactions from other scientists. When I have made the suggestions to concerned citizens groups or the general public, however, the reaction has generally been favorable. This polarization is something which concerns me and which I think should concern the scientific community.

All of the referees were much concerned about false alarms from the early warning system. One suggests: "I also wonder when the public will start to look at these early warnings as unproductive, since it may in one month alone suggest that eating, drinking, and breathing may be exceedingly dangerous for a multitude of new discoveries. Overloading the system will occur based on our knowledge of what papers have appeared in the literature in the last few months requiring action."

Another referee says: "I think our present channels existing today are very accurate to take care of the alertness. As an example, I would like to mention the case of the flame retardants, the saccharin, and the spray paint. . . . I think it would be a waste of money and paper to create a so-called early warning system as described in this paper by Dr. Bross. It probably would have been very accurate to do what he is proposing ten years ago, but I think he is out of date."

A third referee says: ". . . the author can have very little idea of the high proportion of false alarms which occur in the more general field. The early publication of such alarms before adequate evidence is gathered could have serious effect on public confidence in scientists--one can cry 'wolf' too often."

Doubtless many, perhaps most, EHP readers would share this preoccupation with false alarms. From my contacts with concerned citizens, however, I can report that a "serious effect on public confidence in scientists" already exists. It is not, however, a result of scientists crying "wolf" too often. The citizens are concerned because relatively few scientists have been on the side of the public in efforts to obtain adequate protection against environmental hazards. Indeed, the citizens find that much of the time the scientific or medical establishment is on the other side of any effort to circumscribe the uses of technology. For instance, in the recent legislative battles over protecting the public against the potential hazards of recombinant DNA research, the lobbyists against "government interference" included the president of MIT, representatives of Harvard, Stanford, and Cornell, and spokesmen for various scientific societies. So while traditional scientists may give top priority to avoiding false alarms, the public gives top priority to getting protective action on real alarms.

LETTER TO THE EDITOR:
AN ANTI-PUBLIC BIAS

Avoiding "false alarms" at all costs is a luxury that only "pure science" can afford. In "pure science" the usual way to avoid making a wrong statement is to make a nonstatement (but in very technical language). However, to protect the public against environmental hazards, actions are needed and not just empty statements. Taking action always involves non-negligible risks, in part because inaction is itself a course of action. The dread of being accused of giving a false alarm and having one's reputation damaged often results in inaction. It is again a matter of inverted priorities. While reputation is important to a scientist, it is a small personal cost when the health and safety of other human beings is at stake. If factual evidence shows that a hazard exists, I believe a scientist has the responsibility of warning the public irrespective of the opinions of colleagues. Over the years, I have been accused at least a dozen times of sounding false alarms which subsequently turned out to be very real alarms. This has won me no medals from the establishment, but I've saved a large number of lives this way.

The tendency of scientists to talk and temporize indefinitely instead of taking action on a public health issue has consistently worked to the advantage of proponents of hazardous uses of technology and against the public. This may not be a deliberate bias toward technology, but it has the same effect. The Primacy Principle explicitly counterbalances this bias.

The article originally submitted to EHP was accompanied by an example of an early warning report. Readers who might like to see how the format for such a report would differ from a conventional paper will find this example (with some minor amplification) in the May 30, 1977 issue of the *Journal of the American Medical Association*. This early warning report by Bross and Natarajan concerns genetic damage from diagnostic radiation.

The editorial history of this article indicates the need for formats and reviewing procedures that would facilitate the communication of information on environmental hazards to the public. In this instance, the publication was held up about a year. Other warning articles have been held much longer or delayed indefinitely. It is especially difficult for authors to publish their findings if they are not well known, are not connected with a well-known institution, or lack special credentials. However, an early warning report should not be judged by the name of the author, but by the data and methodology.

Although the report by Bross and Natarajan was rejected for a variety of reasons, none of them involved any specific faults in the data, the methodology, or the quantitative conclusions. The objections were to style, format, and trivia and not to substance. One reviewer was so annoyed that I had not cited his negative findings, he cited them in his review. Another reviewer objected that there was no confirmatory report--an unreasonable requirement for an early warning report. Actually there are the independent, mutually confirmatory findings of Mancuso, Stewart, and Kneale which are coming out in *Health Physics*. Lack of comprehensive review of literature, lack of a general discussion of radiation hazards, and lack of other traditional trappings were other reasons for objections. Such irrelevant and persnickety reasons were given for keeping the news about genetic damage from ordinary diagnostic radiation from the public for a year.

There is nothing exceptional about this editorial history of a paper on the hazards of low levels of ionizing radiation except that the delay was so short. Other papers in this area have been blocked by the proponents of radiation technology for years, and in some cases indefinitely. I have given a congressional committee documentary evidence of the partly successful attempt to block publication of the Mancuso, Stewart, and Kneale article (it was delayed until after funding was cut off "for lack of publications") and of the successful attempt to suppress a previous report of cancer deaths among the workers at the Hanford reprocessing plant. A reviewer chides me for my "jaundiced view of scientists," but when this sort of thing has been going on in the radiation area for more than two decades, one gets a jaundiced view.

In part, these efforts to block publication may be a matter of the research area. However, whenever the technological hazards involve many millions of dollars, I have found that the game is played this way. The strategy of the proponents of hazardous technology is to delay or confuse the health issues by any and all means, and they can always find some scientists to help them do this. The same reviewer says "In contrast, the cases that I can recall of delay in dealing with real hazards are few." On my part, I have seen very few instances of prompt and effective action to protect the public against carcinogens from products or processes that were of great economic importance. We still have not dealt effectively with the health hazards of cigarettes, although the facts of the matter were clear 25 years ago when I just became involved.

A strong affirmation of the Primacy Principle and steps to implement it are, I believe, a necessary and useful first step toward giving the public the protection against environmental hazards that it so badly needs. As I told congressional committees on June 14 and June 17 of this year, cancer is a preventable disease, and with effective primary prevention programs we would be well on our way to preventing it by the end of the century. To do this, however, we have to change the approaches and attitudes that have blocked action and progress in recent years and restructure our priorities to favor doing something about environmental hazards instead of just talking about them.

Irwin D. J. Bross
Roswell Park Memorial Institute
Buffalo, New York

WHAT YOU CAN DO

If you feel that your life deserves priority over technology and you would like to see the decisions of the federal regulatory agencies reflect this priority instead of the opposite one, then one immediate step you can take to save your life is to let your representatives know how you feel. Your health should be a top issue in political campaigns. If your representative responds with stereotyped "bedbug" letters or empty double-talk, perhaps you need a new one who will go to bat to protect your health and safety. A resolution laying down the Primacy Principle as a guideline for the regulatory agencies may not get through Congress easily or quickly. However, the voting record will be a very good way for you to tell the congressmen who are on your side from those who are not. In the long run, one of the most important steps that you can take to save your life is to have legislators who realize that their political life will depend on what they do to save your life.

chapter 4

GROUND RULES FOR SCIENTIFIC CRITICISM: WHAT IS FAIR AND WHAT ISN'T

Applying the ground rules in the context of low-level ionizing radiation hazards.

INTRODUCTION

The first part of the NRC presentation ended with the words: "There is certainly more than enough evidence here for federal agencies to take immediate action to protect workers and the public against unnecessary exposures in the 1-rad range." Why, then, was it necessary to continue with the presentation?

There were two reasons: One focused on the political realities of the hearing and the other on the scientific realities. The two are somewhat intermingled. The political reality was that the Nuclear Regulatory Commission, Federal Drug Administration, and other federal regulatory agencies may have strong factual evidence of a serious health hazard presented to them, but often do not take immediate action until it is forced upon them. If there is a real scientific "controversy" or if (as is easy to stage-manage) a fake "controversy" is generated, action to protect the public can be tied up for years or may never be taken. So it was not enough to present the facts and the findings; it was necessary to take preventive measures to minimize the real or spurious "controversy" that was inevitable with such theoretically and practically important results.

Answering the objections even before they could be made reduces their impact. Criticisms that might occur because the methods and procedures and the underlying arguments had not been made clear might be avoided by further clarification. So it was useful to go back and fill in some of the details that had been mentioned without elaboration in the first part. It was also necessary to consider the valid scientific objections that might be made and how they might be met by the design or analysis of the study.

The second part of the NRC presentation, therefore, points out and discusses the legitimate lines of criticism for this class of study and gives more details concerning the data and analysis.

LEGITIMATE LINES OF CRITICISM

> The problem is that such action would be regarded as a threat to the economic interests and prestige of the powerful constituencies of the Big Science federal agencies with regulatory responsibilities. Persons in the nuclear industry, in the health industry and radiology, and in the Sciencegate cover-up will do everything they can to block such action--with some help from persons in the federal agencies. An all-out effort will be made to discredit, suppress, and stop our radiation research. It has already been partly successful. In August of this year I will have to give termination notices to Mr. Nachimuthu Natarajan, Dr. Marcella Ball, and the others on my staff who worked with me in this research.
>
> There has been an all-out effort to discredit Mancuso's findings, and there will be a similar effort to discredit our work. These efforts involve a few--very few--genuine scientific criticisms and a lot of the public relations gimmicks which have been used in the Sciencegate cover-up. If the public is forewarned, I believe it can distinguish its friends from its enemies, honest scientists from double-talking administrators. So I will say a little about the gimmicks that will be used to discredit us and how the Galilean rule can be used to counter them.

The basic strategy of those who create fake "scientific controversies" to block public health action is to convince the public that there are differences of opinion between experts and, therefore, the issue is beyond the understanding of the public. The difference between the new estimate of the doubling dose and the old estimates would be portrayed as differences of opinion. Actually, the new estimate comes directly from the facts--no opinions are involved. We put the question to the data: What doubling dose estimate best fits the facts? We pushed a button and got the answers you saw in a matter of minutes. It took much longer to reach the point where this was possible, months to develop the software, and years the know-how to do the job. However, my opinions about doubling dose could not affect the results. Remember, genuine science does *not* rely on expert opinion. According to the Galilean rule the facts are the final authority.

The three lines of legitimate scientific criticism are specified by the Galilean rule. A critic can focus on the facts, on the theory, or on the machinery for confronting the theory with the facts. The latter machinery is here an automated procedure which might be faulted by going into the computer program. No one has asked to see this material.

The second line of criticism, the facts, come from the Tri-State Survey, and some of them are shown on the next slide [Fig. 5, Slide 8]. It is usually quite easy to fault biomedical data. There are a lot of standard gimmicks for this purpose. However, the Tri-State data is unusually good--for reasons which I'll explain in connection with the findings on childhood leukemia. No data are perfect, but this data is certainly good enough for decisions on public health action. Any critic who denies this should be required to present proof of performance--factual evidence that the critic has done a study which is as good as the Tri-State. None of our past critics could present such evidence of competence.

The third line of criticism would focus on the theory. In these slides [Fig. 6, Slide 10] the theory comes into the calculations of the numbers called "expectations." In ordinary language the distinctions between theories and opinions or between hypotheses and speculations are often blurred. In the sciences, the Galilean rule makes a very clear distinction between a verified theory and an unsupported opinion. In my presentation, speculations are identified as such and all of the theory has been validated by the Tri-State facts.

GOOD AND BAD ANALYSIS

In our studies, the theory is expressed both in ordinary language and in mathematical languages. I'll have more to say about the mathematical model when I come to the findings on childhood leukemia where the model is similar but a little simpler. Let me emphasize, however, that the scientific validity of a theory does not depend on the language in which it is expressed. Validity depends on how well the theory describes or predicts events in the world about us.

At this point I will describe the theory in ordinary language and merely mention the mathematical translation. The basic ideas in good scientific theories are usually simple, and this theory is no exception. The idea here is that when human beings are exposed to low-level ionizing radiation, a relatively small proportion suffer any lasting effects. This is what gives the misimpression that the radiation is harmless. However, those who are affected will have a greatly increased risk of getting leukemia or certain other diseases after a period of years. To express this idea algebraically, we introduce three abstract quantities called "parameters." Later we can get numerical estimates of these parameters from the data.

For the adults, the parameters and their estimates are the "proportion affected" which is about 1 in 20 at the lowest dosages, the "relative risk of leukemia" which is about 10, and the "relative risk of heart disease" which is about 3 This tells us that while relatively few are affected, those who are affected have greatly increased risks of diseases. Notice that when this theory is expressed in plain English there is nothing very surprising or difficult or even controversial about it. And it has been carefully validated.

Before I turn to the effects of low-level radiation of the children of persons exposed, there are three brief points I want to make. The first is clear from the present slide [Fig. 6, Slide 10] and that is that we are dealing here with what is a relatively large series of cases for a disease as rare as leukemia. We are not dealing with the 5 or 10 cases that are often reported, but with a series of more than 2000 leukemia cases and about as many persons who were randomly sampled from the three-state area. These are very solid facts.

My second point is made by the formula on the next slide [Fig. 7, Slide 9] which we used in calculating dosages. All I'll say about dosimetry at this point is that it always opens a can of worms and you can fuss about any numbers used. Fortunately, the shape of the dosage response curve and our findings are not greatly affected; different numbers would mainly result in relabeling the axis, not changing the picture.

My third point concerns computers. As you've seen, we make heavy use of computers, but this doesn't add or subtract from the scientific validity of our findings. Computers are a tool which can be used foolishly or wisely, but this depends on the user and not on the tool. As a professional biostatistician for 25 years, I am well aware that it is hard for most people, and most scientists, to appreciate the distinction between a good statistical analysis and a bad one.

It is not hard to do a bad analysis of our data or of Mancuso's data. Indeed, several have already been done. All that is needed for a bad analysis is a canned computer program, say, an ineffectual, general purpose method such as the Mantel-Haenszel test, and a lack of competence in biostatistics. These resources are widely available. Then all you have to do is throw the data in a hopper, push a button, and come out with a lot of computer printout. It is unnecessary to spend a single minute in thinking about the scientific context of the data or to know anything about biostatistics. For those who want to do a so-called "independent analysis" in order to discredit positive findings, there is a very nice feature of incompetent statistical analyses; namely, they will tend to produce shadowy or negative findings even though a good analysis will come out with strong positive findings.

A good example is the report that Kneale just gave in Vienna on the Mancuso data. In addition to his analysis which gives clearcut positive findings, he did a Mantel-Haenszel test on the same data. It gives at best borderline significance at the 5 percent level. An incompetent analysis of our data on x-ray during pregnancy appeared in an issue of the *British Medical Journal* and managed to miss the strong radiation effects in this data.

THE TRI-STATE SURVEY

The original order of presentation was reversed because I wanted to be sure I could present our new findings on adult leukemia. Now, as time allows, I will present our work on childhood leukemia and spend more time on the details of data and the theory.

Let me start by telling you a little more about the famous Tri-State Survey. This is a survey which was carried out in sample areas of upstate New York, Maryland, and Minnesota between 1959 and 1962. The main purpose was to confirm or disconfirm the findings of Dr. Alice Stewart that the radiation delivered in pelvimetry could produce leukemia in the child. At that time, this finding was very controversial.

The practical difficulty in studying a relatively rare disease like leukemia is that an enormous population base is needed to get sufficient cases for a detailed analysis. Here the effort was made to get all of the cases of leukemia in a population base of 13 million persons over a three-year period.

To obtain a valid control series, a stratified random sample of households from the same area was drawn--as statistician I was involved in this aspect. Interviewers were then sent out with a long and detailed question schedule to interview a respondent in the households of the leukemia patients and of the random sample controls.

This study is one of the few epidemiological studies that was done "double-blind." In other words, persons being interviewed did not know that they were being interviewed about leukemia, and the persons doing the interviewing did not know whether they were interviewing in a household from the random sample or with leukemia. The key questions that might have helped the interviewer to guess the status of the household were kept to the end of the schedule. There are few surveys conducted today which are nearly as good as this one. That this study was so many years ahead of it's time is due to the foresight and competence of Dr. Abraham Lilienfeld who conceived it and set it up. So we are starting with the best data of its kind that is currently available.

Here is an example of the data that comes out of the Tri-State Survey. In the slide you see Table 2 [Fig. 8]. We will see Table 1 [Fig. 9A and B] in a moment. This slide gives the observed and expected counts in the children 1 to 4 four years of age. So let me take a little time to tell you what the numbers in this simple table mean, because you will be seeing more complicated tables like it. There are two series shown, one marked "leukemia" and one marked "no leukemia." The "no leukemia" children are from the random sample controls that are used in this survey as representatives of the general population. Now there are two categories for the indicator disease: "no" and "yes." This means "no indicator disease" or "yes, an indicator disease was reported." In other words, where there is a yes answer, the child had one or more of the following childhood diseases: asthma, urticaria, eczema, pneumonia, dysentery, or rheumatic fever.

The numbers shown as observed numbers are the actual counts of the children in the Tri-State Survey. For instance, there were 27 cases observed in the interviews of leukemia households where there was no report of an indicator disease, and only three where such a disease was reported. What we are looking at in this table is the children where there was *no* report of an x-ray exposure to either parent or to the child. The numbers called "expected," which you see directly below the 27 as 27.7, are the numbers predicted by the mathematical model which describes how the Tri-State Survey was carried out and what the various risks of diseases are. The model simply gives in mathematical language an accurate description of what went on in the Tri-State Survey. Since there is no radiation reported here, the model does not involve radiation hazards.

We can now see whether the model fits the facts. To do this we compare observed and expected. What you can see very clearly from this table is that the observations and the expectations are very similar. Therefore, we can say the theory fits the facts. If we want to get fussy, we can calculate a goodness-of-fit statistic.

WHY STUDY CO-OCCURRENCE?

You might wonder: Why are you looking at all those other childhood diseases when you are mainly concerned with the risk of leukemia from radiation in the 1-rad range? The answer involves a basic scientific problem that underlies most of the technical problems here. The problem is: Among the leukemia cases there will be some cases that are produced by the radiation. But there are going to be many other cases that are produced by extraneous factors in the environment--possibly background radiation, possibly chemical exposures, or possibly still other variables. We would like to have some way to distinguish those cases which are due to the radiation from those cases which are not. Unfortunately, we can't exactly do this. What we can do is take a step in this direction by finding other diseases with the same cause. The breakthrough here came when we found some clues that the radiation produced both leukemia and the indicator diseases.

What this means in practice is that if we focused our attention on the relatively small number of persons who had reports of certain other diseases as well as leukemia, then the cases we were looking at were much more likely to be produced by the radiation than those in the rest of the series. In other words, this provided an enriched series where, in a probability sense (and I emphasize only in a probability sense), we are dealing with persons who were more likely to have had their leukemia produced by the diagnostic radiation. In the remainder of the series, the cases would be relatively more likely to have their leukemia produced by other factors.

As far as the general public is concerned, I don't think it is difficult for anyone to see this point: If radiation produces leukemia *and* certain childhood diseases, then we are going to be able to see the effects of radiation most clearly if we look in a series where the individuals were given diagnostic radiation and where there was a report of the childhood diseases as well as the leukemia.

To exploit this scientific finesse we had to develop a mathematical model. To show you what a mathematical model looks like, and where the "expected numbers" come from, I'm going to flash another slide [Fig. 9A and B, Slide 3]. This shows the formal statement of the model in mathematical language. The mathematical details of the algebra need not concern us. Whenever there are questions about algebraic operations, these are always resolvable in principle (and usually in practice) simply by doing a series of arithmetic computations. In other words, this is not a matter of opinion; it is simply a matter of going through the calculations. People can agree reasonably well when this is all that is involved.

The details of the model are not particularly important. One point to note is the distinction between the basic model and the adjusted model. The first does not involve radiation and the second brings in the hypothesis about radiation that we want to test.

There are always technical difficulties with models. Coping with their problems is largely a matter of experience with models. We have 20 or 30 publications dealing with mathematical models applied to actual cancer situations--breast cancer, leukemia, various other situations. We know what we are doing when we deal with models. To put it bluntly, not many other people can do this job. Most mathematicians don't know enough about biology and about cancer to create this kind of mathematical model. Most biologists and epidemiologists don't know enough about mathematics to do it.

The hypothesis which I have mentioned, namely, that radiation can produce diseases of childhood in addition to leukemia, is something which we can express in the mathematical language fairly easily. In the "adjusted model" at the bottom, the radiation hypothesis is expressed in algebraic language.

Now some people find it hard to accept the notion that the same factor can produce more than one disease, something which tied up the cigarette-cancer issue for a long time. Nowadays we know of many environmental factors which can produce a multiplicity of diseases. Again, some folks like to quibble about the assumptions of a mathematical model. But these quibbles aren't science. There is a rule in science, the Galilean rule, for settling such matters.

DOES THE THEORY FIT THE FACTS?

Does our theory fit the facts? The next slide [Fig. 10, Slide 4] shows how we can answer this question precisely when we express the theory and the facts precisely. We've seen in the first slide [Fig. 8, Slide 2] that the basic model fits the facts very nicely when there was no report of radiation. When there is a report of radiation to mother and fetus, something new has been added. Table 3 [Fig. 10] shows the situation in the case where the x-ray was delivered to the mother and fetus.

Again you can see immediately, without going into the technicalities, that the model that worked in the previous slide [Fig. 8, Slide 2] fails to fit the facts in the table on the left [Fig. 10]. However, the table on the right for the "adjusted model" fits quite well [Fig. 10]. The table on the left assumes there is no radiation effect and we have to reject this assumption. The table on the right is the "adjusted model" where a specific hypothesis about the nature of the radiation hazard is used in the calculations. This hypothesis is confirmed by the good fit to the facts.

Whether this hypothesis is stated in plain English or a mathematical language, what the model assumes is this: There is a certain part of the population that is affected by exposure to low-level radiation, and the rest of the population is not affected by the radiation. The distinction between the affected and unaffected group is that in the unaffected group the risks are the same as before (as in Fig. 8) for both leukemia and the indicator diseases. On the other hand, if an individual is affected, then the risks of the leukemia and the risks of the indicator diseases will both be markedly increased.

The three parameters here are: (1) The proportion of individuals affected by the radiation which is 1 percent for the children; (2) the relative risk of the leukemia in the affected series (which is 50 for the children--the risk is 50 times greater than it would be in an unaffected individual); and (3) the relative risk for the indicator diseases (which is five for the children--meaning that the risk is increased fivefold). Now when we use this adjusted model in the calculations then we get the expected numbers shown under the heading "adjusted model."

There is a working rule called Occams' Razor, which is a medieval name for the theological principle "Never multiply entities without necessity." We don't introduce a whole new set of parameters unless we really need them. Now, do we need them here? This we can see by looking at Table 3 [Fig. 10] and the relationship between observed and expected in the model and then in the adjusted model. If we look at the situation for the original model, we see that in the category of indicator diseases there are 12 cases observed, but only 3.4 expected. So you are looking at a model that fails to fit the facts. If we want quantitative evidence of this we can calculate chi-square, and this tells us that the model fails. When you look at the corresponding entries for the new or adjusted model you see that the numbers are 12 and 12.0. The numbers don't fit perfectly in all the cells, but basically what we have here is a model which *does* fit the facts. It is not necessarily the only model that will do this, but it is a model that fits the facts. This means that we have a scientific reason to prefer the adjusted model to the original model.

In this way, we confirm the hypothesis of the model that radiation produces genetic damage that will be expressed in the child as indicator diseases and leukemia.

The task of the scientist doesn't stop with the single 2 × 2 table of Table 3 [Fig. 10]. We need to look at other age groups or radiation exposures, and this is what we have done. For example, in this particular body of data, we can look at the situation for all of the age groups in the next table [Fig. 11]. What you can see from Table 4 [Fig. 11] is that the observed and expected values aren't in perfect agreement. They should not be perfect because there is always sampling variation. This is what chi-square is all about. However, you can see directly that the expected numbers are similar to the observations in all of the cells of Table 4 [Fig. 11]. Even more important, the expected values have the same patterns as the observations. In the radiated group there are always more children where both indicator diseases and leukemia are found than would be expected if there was no radiation damage. On the other hand, in the data where there is no radiation delivered, the original model fits rather well and is validated as a description of the Tri-State Survey data in the absence of radiation. The more complicated "adjusted model" involving radiation effects is validated by data that does involve radiation exposures.

chapter 5

THE GALILEAN RULE

How the basic rule of modern science can be applied to the development of a valid scientific argument.

THE PRIME DIRECTIVE

Of all the scientific strategies to save your life, the most important single one is the Galilean rule. It is also the prime directive of modern science:

The Galilean rule: A theory must fit the facts.

Modern science began about four centuries ago when Galileo used this strategy to demolish the Aristotelian theories of motion that had stood unquestioned for two millennia. This was an essential first step in the development of modern physics. It also got Galileo into trouble with the Catholic Inquisition.

To explain this strategy, its strengths, its weaknesses, and its application to a public health issue such as the hazards of x-rays, a question-and-answer format will be used. This kind of dialogue can bring out the key points somewhat more clearly than a narrative text.

question: Why is it necessary to have this chapter? Isn't the Galilean rule simple and straightforward enough for anyone to understand immediately?

answer: The basic idea of the Galilean rule is simple and clear. But when we try to apply this idea in an actual scientific study, a number of questions and complications are likely to arise. These may be real problems or, as in the critique and questions at the NRC session, they may be an effort to produce confusion and controversy that will cloud the issues. Knowing more about the Galilean rule and its application can help to distinguish real science from pseudo-science.

question: What kind of questions and complications are you talking about?

answer: There are various kinds. Some concern the three components of the Galilean rule: the theory, the facts, and the "fit" between the theory and facts. The questions can be specific: "What are the facts in a study, and are they relevant and reliable?", or the questions can be general or philosophical: "What are facts?" Indeed, since the Galilean rule is the yardstick for "truth" in the sciences, one might ask: "Is this a proper yardstick?", or even "What is truth?"

question: Do we have to answer all of these questions to understand the Galilean rule? Hasn't that last question been debated for over 2000 years without settling the matter?

answer: No, we don't have to answer all of these questions. However, you may need to cope with these questions if they are raised. For example, "What is truth?" is a very tricky question because the meaning of a word such as "truth" depends on the usage in a particular language. For instance, the usage in the languages used in religions will be different from that in the languages used in the sciences. Fortunately, we are concerned here with scientific languages and with ordinary language. We don't have to get bogged down in all the linguistic quagmires of philosophy. We do, however, need to know how to stay out of these quagmires.

question: Aren't you getting into a vicious circle here? Aren't you making the Galilean rule the criterion for truth in science?

answer: No. The Galilean rule characterizes the usage of scientific languages in first-rate studies. In other words, such studies scrupulously follow the rule that a theory must fit the facts. But this isn't something that I impose or anyone could impose on the speech and behavior of scientists. The Galilean rule is not the prime directive of modern science in the sense that it is imposed from above. Rather, it is a working rule that all good scientists habitually follow on their own volition.

question: Is this usage of "truth" in scientific languages different from the ordinary usage of the word?

answer: There are differences in usage, but these are not basic differences from ordinary language. A "true" report of events is a statement which corresponds with what actually took place. In science, the report may be called "data." Nevertheless, the criterion for "truth" is the same. A "theory" in science corresponds to a "summary" or "general" statement in ordinary language. A general statement about a set of reports would be called "true" in ordinary language if it is concordant with those reports. The Galilean rule requires a similar concordance for a scientific theory--it should not be contrary to the actual data. There are differences in usage, but they are not fundamental. For instance, a greater accuracy and precision of statement may be required in the sciences. This, however, is only a matter of degree.

SCIENTIFIC THEORIES

question: Aren't you glossing over differences? Isn't a scientific theory something entirely different from any ordinary, general statement?

answer: No. There is a mystique about scientific theories that exalts them. However, this is not really science but merely the kind of self-glorification that goes on in most enterprises and professions. In what is traditionally regarded as the first application of the Galilean rule--the quasi-legendary dropping of two cannonballs from the Leaning Tower of Pisa--the two theories were as follows:

Aristotle: Heavier objects fall faster than lighter ones.

Galileo: Heavier objects and lighter ones fall equally fast.

Statements of this kind are not uncommon in ordinary discourse.

question: But that was at the beginning of modern science. What about Einstein's theory of relativity? Surely that wasn't something that is like an ordinary statement, is it?

answer: Relativity theory is expressed in an algebraic language. It involves statements which are more complex, abstract, or sophisticated than most statements in ordinary language. However, these are not fundamental differences, but only differences in degree. Einstein's theory was an attempt to describe events in the world about us. It was duly tested by the Galilean rule. The first critical test on actual data involved photographs taken during an eclipse of the sun. If at some point Einstein's theory fails to fit the facts, it will suffer the same fate as the Newtonian theory it replaced. Physical theory has evolved in the 400 years since Galileo, but this is a change in content, not a change in character.

question: But certainly you must admit that the relativity theory isn't something ordinary--that it took an Einstein to discover it. Doesn't that make it fundamentally different from ordinary statements?

answer: No. This question confuses two different contexts which the late Dr. Hans Reichenbach called "the context of discovery" and "the context of justification." In the context of discovery, it is certainly true that an Einstein is unique. The discovery of the theory of relativity required a creativity and competence that was far out of the ordinary. The creative leaps in the context of discovery are not reducible to logic or common sense. Einstein is unique in the same way that Shakespeare or Goethe or Tolstoy is unique. But what concerns us here is a different context--the context of justification.

question: Are you saying that *in the context of justification*, science is reducible to calculation and common sense?

answer: Exactly. No matter how a theory is conceived or created--whether it is the product of clear thinking or foggy mysticism--the theory must eventually be stated in a form which can be put to the Galilean test. Again, no matter whether the theory is put forth by a VIP in Big Science or by the lowliest graduate student, the theory is rejected if it fails this test. In the context of justification, all theories stand or fall on whether they fit the facts. For all its elitist tendencies, there is this basic democratic principle in the scientific enterprise.

LIMITATIONS OF SCOPE

question: You mentioned limitations on the scope of the Galilean rule. Just what are they?

answer: There are some scientific issues where the Galilean rule alone resolves the matter. There are other issues where the Galilean rule is used indirectly and in conjunction with other rules. You shouldn't expect one simple rule to do everything.

question: What sort of issues are directly resolved?

answer: What the Galilean rule can do is to reject a false theory--a theory that doesn't fit the facts. Even here, the rejection may be decisive--there may be almost no chance that the theory is correct--but it is not absolute. Thus, given the Tri-State data, the null hypothesis (i.e., no association of leukemia and other diseases with radiation) can be decisively rejected. Along with this hypothesis, we can reject the myth that low-level radiation is harmless. But while the Galilean rule can be used directly to destroy an erroneous hypothesis or dogma, it can only be used indirectly in establishing a new theory to replace it.

question: You seem to be saying that the Galilean rule can kill a theory, but it can't prove a theory. Why is this?

answer: This question is not an easy one to answer, and a full answer gets into some technical aspects of linguistics. The key word here is "uniqueness." The crux of the problem is that we can describe the world in more than one way and still give a fairly realistic account of what is going on. So when a theory fits the facts, it is not necessarily the only theory that can fit the facts. It is not unique in this respect.

question: Does a theory have to be unique?

answer: Yes and no. If one is talking about "proof" in an absolute sense, it would have to be unique. If one is speaking in a realistic or practical sense, no scientific theory is ever established absolutely, and it is always provisional in this sense. It is the best description of the world that is available at a given time, but may always be replaced by a better description.

question: What you seem to be saying is that the Galilean rule can't lead to final conclusions. Doesn't this leave everything up in the air on health issues?

answer: Not really. With the help of the Galilean rule and some further strategies related to it, science approaches the truth by what is called a "sequential process." In other words, it gets closer and closer to the truth step by step. Even though we don't know the final conclusion, the steps take us to a point where we have very reliable information. The theory fits the facts well enough to give us estimates of the probabilities we need to weigh the advantages and disadvantages of public health action. The Galilean rule may not take us all of the way to the truth, but it can bring us close enough for practical purposes.

question: Isn't this just an alibi? Don't you need to be sure a theory is correct before you take action based upon it? Aren't you admitting that we can't be sure an epidemiological theory is correct and, therefore, mustn't we wait until we are sure?

answer: This is the line of argument which has often been used to block action to protect the public. It relies on the myth that science gives certain or infallible results, a myth believed by some physical scientists and engineers. But this isn't science--this is hokum. The limitations of the Galilean rule apply equally to the physical sciences or to biomedical sciences such as epidemiology. Whether calculating orbits of space vehicles or risks of radiation, we are always dealing with probabilities of events in the world about us, not certainties. If we wait for certainties, we postpone action indefinitely. Technologists may like this idea when it applies to postponing restrictions on technology. However, they would scream bloody murder if this argument were applied the other way. Suppose we said: There can be no further deployment of any radiation technology until we can be absolutely certain that no one is going to be hurt by it. There could be no further deployment of technology.

THEORY AND ACTION

question: But I still don't see how we can act on a theory until we know it is correct. We don't want to act on a false theory, do we?

answer: No, we don't want to act on a completely false theory. However, the Galilean rule does allow us to reject theories that are outright false. The difficulty arises in distinguishing between theories that are nearly correct. For practical purposes, a theory that is nearly correct may be more useful than one which is theoretically more accurate. Thus, even in the Space Age, the arithmetic for calculating orbits generally follows the theory of Newton and not of Einstein. Although Einstein's theory fits the facts better when velocities approach the speed of light, for attainable velocities the numerical difference between the theories is negligible. The newtonian arithmetic is much easier. So the possibility of a better theory is no bar to decisions and actions based on a good theory. Physicists should know this from the history of physics, but they don't seem to see this point.

question: You seem to be saying that the co-occurrence theory which you've tested may someday be replaced by a better theory, but this might not matter. The new theory would have much the same implications for action--say, cutting the permissible level--as the theory under test. Is that right?

answer: Yes. The public should realize that anyone who claims infallibility for science cannot be a genuine scientist. When you work with actual data, you very soon learn about the fallibility of science. In drawing conclusions or taking action in science, we should do the best we can. We shouldn't kid ourselves or others that we are perfect.

question: Don't you lose something if you aim at less than perfection?

answer: What you lose is an illusion--an illusion of perfection in an imperfect world. Ideal or unattainable standards are an obstacle to developing realistic criteria. Giving up perfection doesn't mean that we can't discriminate between good and bad theories. The Galilean rule does this. The theory that low-level radiation is harmless is in flat contradiction to the Tri-State facts. We have to admit, however, that we cannot always discriminate between theories which are concordant with the facts. Even here, however, there are auxiliary rules that give some basis for preferring one theory to another when both fit the facts.

OCCAM'S RAZOR

question: What are these rules?

answer: One is called "Occam's Razor." It dates back to the days of scholastic philosophy when the highly publicized controversies were theological rather than scientific. To avoid the foolish controversies that were giving theology a bad reputation (as in Rabelais), William of Occam (England) proposed the following rule:

Occam's Razor: Never multiply entities without necessity.

This idea appears in modern science in various paraphrases. One of these is called the Principle of Simplicity. This might be phrased as: Choose the simplest hypothesis that fits the facts. This rule establishes a preference system when there is more than one hypothesis that fits the facts.

question: Why should we prefer the simplest hypothesis in a world which we know is a very complicated place? Should we suppose instead that the world is simple?

answer: No. The preference is a matter of language, not of fact. *We give the simplest hypothesis pride of place because this makes the sequential process for getting to the truth more efficient.* Moreover, for practical purposes, the simplest hypothesis is likely to be the easier to communicate and to use. The rule is sometimes misinterpreted as a statement about the world, but then it would be foolish. Instead, we have to recognize that any theory, including a theory in algebraic languages, is always going to be an oversimplified description of a complex reality. We prefer the simplest such description because it is the easiest one to reject by the Galilean rule. This way we are led, step by step, to more complicated descriptions. However, we can avoid unnecessary complications.

question: How is Occam's Razor used in a scientific argument?

answer: The Razor is mainly used to cut through the tangle of hypotheses and counterhypotheses that so often occurs in health issues. In effect, it gives a priority system for bringing hypotheses up for testing. You start with the simplest hypotheses. Then you work your way into increasingly complex ones. Eventually, you get to a theory that fits the facts--if you are lucky.

question: What do you mean by the "simplest" hypothesis?

answer: This can be quite a complicated question. Sometimes simplicity is straightforward. For instance, the mathematical models used in the NRC presentation have a number of unknown quantities or parameters. The simplest model would set all of the parameters at a value corresponding to the hypothesis of "no relationship" (e.g., relative risks at unity and proportion affected equal to zero). The more parameters that require non-null values in order to fit the facts, the more complicated the model. In mathematical languages, simplicity may be fairly simple. When the real world intrudes, this is no longer the case.

question: Why not?

answer: When we get into the scientific content of a theory rather than its formal mathematical structure, it would not always be true that a model with three parameters is necessarily more complex than one with two parameters. The nature of the parameters may be quite different in the scientific context. For instance, consider two rival theories (or one-parameter mathematical models) that fit a given body of data about equally well. If in the first model the parameter represented experimental error (say, an estimate of variance), and in the second a radiation effect, the first model would be preferred.

ORDERING BY SIMPLICITY

question: How could you say the first model was simpler by Occam's Razor?

answer: The argument would be that we know that there is always experimental error or sampling variation in a study. In this sense, the first parameter is necessary. However, if the results can be explained simply in terms of the inherent variation, there is no need to invoke a new entity--a radiation effect. So we don't introduce a new entity without necessity.

question: Are hypotheses always clearly ordered with respect to their simplicity?

answer: No. There are orderings that are generally accepted in a given science, but sometimes the order isn't clear. But this is not a grave problem. It mainly affects the sequence in which the hypotheses are tested. In general, however, scientific hypotheses and counterhypotheses fall into four broad classes, and Occam's Razor orders these classes.

question: What are the four classes?

answer: First, there are null hypotheses which say, in effect, that there is nothing in the data that cannot be explained by the inherent experimental error in the study. In other words, the numbers are simply random ones. Since there is always experimental error, this is the simplest class of hypotheses.

Second, there are hypotheses based on artifacts. These say, in effect, that there is nothing in the data except what the scientist puts into it by the way he or she did the experiment or survey. For example, a bugaboo of any survey is "interview bias." A good survey takes precautions in the design and conduct of the survey to avoid or minimize such biases.

This second class of hypotheses are generally raised as counterhypotheses--alternative explanations to the theory under test. Despite a general belief to the contrary, in my own experience the problems with artifacts tend to be much worse in laboratory experiments than in field surveys. The reason for this is that the laboratory scientist tries to control his or her experimental error by direct intervention. In manipulating the material to achieve a desired effect, he or she may introduce artifacts.

question: Is the "next-to-simplest" ordering of hypotheses based on artifacts due to the fact that artifacts are almost as ubiquitous as experimental error?

answer: Yes. This is why an expert critic looks first at the Methods and Procedures section of a scientific paper, while the usual critics--who are not very competent--look first to see whether or not they like the results that are reported. Artifacts can be subtle, and some Nobel prizes have been awarded for results later found to be spurious and artifactual.

question: What is the next class of hypotheses?

answer: These are hypotheses based on well-known real effects. The nature of this class differs from one science to another, but each science has its list. In epidemiological research, for instance, "demographic" effects may explain the results. For instance, cancer risks go up with age so it is often important to deal--as in the analyses presented here--with separate age groups. There may also be important sex differences. We can avoid being misled by such effects by simple means--analyzing the sexes separately--or more complicated adjustments. By Occam's Razor we don't want to introduce a theory involving new radiation effects if we can explain the data by factors we already know about.

question: What is the fourth class of hypotheses?

answer: These are hypotheses that we want to test which are not known or established factors. For instance, hypotheses of occurrence or co-occurrence of disease as a consequence of exposure to radiation would be in this class.

question: But aren't radiation effects well known?

answer: At high-dose exposures, yes. But here we were looking at exposures to the low-level radiation of diagnostic x-rays. In general, a substantive hypothesis under test would go into this category even if there was good evidence for it. The point is that the simpler hypotheses would first have to be eliminated before a more complex substantive hypothesis would be preferred.

THE PROCESS OF ELIMINATION

question: Does this step-by-step process of elimination get to the truth of the matter?

answer: Well, this is the slow-but-sure sequential process that modern science uses in getting as close as possible to the truth--in getting to theories which are reliable for practical applications such as public health action. The Galilean rule and Occam's Razor can be combined into a powerful mill that grinds slowly, but eventually grinds fine. There is no absolute truth, but one can get relatively close to the truth.

question: The process of elimination that you have described is certainly a lot more complicated than the simple Galilean rule that you started out with. It sounds as if establishing a new scientific theory in this way is a very long, slow, and difficult process. Is this why it is so hard for a scientific theory to become accepted?

answer: No. The delays and discouragements in the development of new scientific theories are mostly due to the fact that science is a human enterprise, and humans react to a challenging new theory in human terms rather than in formal terms. Neither factual evidence nor formal processes of elimination can compel a scientist to accept a theory that poses some kind of threat to him--to his prestige, to his position, to the institution or corporation that he works for. The process of elimination of competing counterhypotheses can be difficult and time consuming, but it is not nearly as much of a hurdle as getting other scientists to understand--or even listen to--a disturbing new theory.

question: If scientists act in such highly personal and subjective ways, how does any new theory gain acceptance?

answer: In two ways. The first road to acceptance is that there are some in the scientific community who are willing to listen to the facts and the formal arguments. Usually, there are the younger persons who have less of a vested interest in particular theories. Occasionally, however, wise old persons can also be open-minded. This may give the new theory a foothold. Then, the second road to acceptance may be opened. This is a process of attrition where the adamant opponents of the theory eventually die off.

question: That process of attrition sounds rather cynical. Surely it is an exaggeration, isn't it?

answer: Perhaps. However, it is not much of an exaggeration. For instance, when I first became involved in the cigarette-cancer issue more than a quarter of a century ago, the sides were drawn early in the 1950s. Despite constantly mounting evidence of cigarette hazards, very few of those who originally denied the hazard ever changed their minds (or said so publicly). By now, however, the leaders of the opposition have died or disappeared from the scene. While the tobacco industry was always able to get some new recruits, these play-for-pay replacements lacked status. Unfortunately, the process of attrition took 15 or 20 years, and the efforts to protect the public were tied up in knots during the period.

question: Why do you emphasize the irrationality of the behavior of scientists in a chapter on the scientific process for getting to the truth?

answer: Because the public has to realize that there is often a wide gap in the sciences and in other professions between theory and practice. It is pointless, for instance, to postpone protection against public health hazards until scientists can agree on what to do. The goal is not to convince recalcitrant scientists because there will never be complete agreement on these issues. What matters is what the facts tell us and not "expert opinion" about these issues. Prolonged debates between the scientist-lobbyists for the polluting industries and the scientist-bureaucrats in federal regulatory agencies represent a tacit coalition of professionals against the public. The lobbyists get salary increases and the bureaucrats manage to avoid taking any action that might be hazardous to their health (fiscal). However, none of this posing and posturing has much to do with the scientific procedures for getting to the truth.

question: Why do you bring up this process of elimination for getting to the truth in the sciences if, in fact, relatively few persons understand and use it?

answer: The public ought to know that there is a systematic process in the context of justification in science, and that this is what first-rate scientists use to establish their theories. In principle, therefore, there is a systematic way to settle the so-called "controversies" in the health area. So the public should realize that the reason these "controversies" drag on for so long in the scientific community is that the game is not being played by the rules.

question: Do you think this is wrong?

answer: While they may be called "scientists" by the mass media, these individuals are not acting as scientists in the public interest. It is not necessarily wrong for these people to act in their own self-interests to promote professional interests, or to act as spokesmen for their agencies, institutions, or corporations. It is deceptive and wrong is to do so in the name of science. It is cheating the public to speak as a scientist and not to follow the basic rules of the enterprise.

chapter 6

SCIENTIFIC AND PUBLIC HEALTH IMPLICATIONS

How scientific findings should be used to protect workers and the public from radiation hazards.

INTRODUCTION

The second part of the NRC presentation in Chapter 4 presented the ground rules for scientific criticism. This was, in effect, a self-critique which evaluated the evidence presented by applying these ground rules for legitimate scientific criticism. This kind of self-criticism is what tends to distinguish the research of good scientists from that of second-raters. This is not an unusual step in the research process, although it is a little unusual to deal with it as explicitly as was done here.

The key question in any genuine critique is: Does the theory fit the facts? After going into some detail on this point, the second part of the NRC presentation concluded with the statement:

> In the radiated group there are always more children where both indicator diseases and leukemia are found than would be expected if there was no radiation damage. On the other hand, in the data where there is no radiation delivered, the original model fits rather well and is validated as a description of the Tri-State Survey data in the absence of radiation. The more complicated "adjusted model" involving radiation effects is validated by data that does involve radiation exposures.

Readers who were convinced that science was a completely logical process might expect that once a clear hazard to humans had been established for dosages of ionizing radiation in the 1-rad range, the regulatory agencies responsible for protecting the workers and the public would take immediate steps to carry out their responsibilities. However, a reader with some experience with federal agencies would realize that the Nuclear Regulatory Commission does not operate in this logical way. Although the NRC might have immediately initiated adjudicatory hearings to cut the permissible levels from the 5 rads that had been shown to produce a serious increase in human risks to, say, the 0.5-rad level, it took no such action. At the very least it should have scheduled the adjudicatory hearing that was promised. At present, however, there is no sign that the NRC even intends to do that.

Anticipating this probable inaction by the NRC, I asked myself: What possible excuse can be given for this failure to take a clearly indicated action to protect the public? Some possible excuses might be that (1) the action implication of the scientific findings had not been spelled out; (2) the details of the biostatistical technology (such as the mathematical model) were somehow unclear; (3) there are questions about the protocol of the Tri-State Survey; and so forth. Of course, if the NRC were really hung up on any of these points, they only had to pick up the telephone to get the full information (or to write a letter if this was more expedient). They never did. I had been ready and willing to respond to such requests for information from Dr. Rothman. (I know that if I had been preparing a critique of this material, I would certainly have wanted to get some additional details). Surprisingly enough, there never was any such request.

This final section of the NRC presentation takes away the only legitimate excuses the NRC might have for failing to do its job.

MATHEMATICAL VS. VERBAL LANGUAGE

Let me emphasize one point which is sure to come up, and in a way illustrates the advantages of having everything formulated precisely in a mathematical language as opposed to informal statements in a verbal language. Since it's just about impossible to honestly deny that there is a strong relationship to radiation showing in this data, the main line of defense for those who want to argue that there is no radiation effect is to claim that there is some other explanation. In technical terms this is called a counterhypothesis. The nature of the counterhypothesis varies, but the general idea would be that, yes, there is a relationship between indicator diseases and leukemia, but this comes because indicator diseases are in some sense precursors of the leukemia. This is the kind of issue which can tie us up in knots if we don't have a clear understanding of what is going on.

If you look at the series where there is no radiation, then you can see that the observations and the expectations are in good agreement. This particular model involves the assumption of independence. Now what does that mean? That means that we assume that there is in fact no correlation between the indicator diseases and the leukemia in a situation where there is no common cause, that is to say in a situation where there is no radiation.

Suppose there really was a correlation between the diseases themselves. Then looking at the totals in Table 4 [Fig. 11] for the series with no report of radiation you ought to see *more* cases than expected of indicator diseases and leukemia. As you see, there are *less*. There are 7 cases observed, compared with 8.5 expected. The counterhypothesis doesn't fit the facts. If it was correct, there should have been 15 or 20 cases instead of only 7. So the counterhypothesis of hypothetical correlation is refuted by the data.

It is important not to get mixed up in technicalities here. It is not a matter of ruling out absolutely any and all counterhypotheses. This might be necessary if we wanted to show beyond a shadow of a doubt that the explanation that we are offering is the only explanation. However, for a prima facie case, this is not necessary. In practice in public health, it's impossible. You can never really rule out all hypotheses that might be raised simply because anyone with any ingenuity can always raise a new one. But anyone who claimed to have refuted our findings by simply raising this counterhypothesis would, for practical purposes, be lying to the public in a technical language. Anyone seriously proposing a counterhypothesis must match it against our hypothesis and show that it can do substantially better. Otherwise, by Occam's Razor this simple and straightforward model would have priority. None of our adversaries have done this so far, and I don't think we are going to see it today either.

DEALING WITH COUNTERHYPOTHESES

Without trying to go into all the details of the data for the mothers who received radiation prior to conception or for the fathers who received radiation prior to conception, I will simply show you two more slides of the situation with the model and the data for these further cases. In Slide 6 [Fig. 12] you can see that the theory fits the facts nicely when we are looking at preconception radiation to the mother, in this case a very purified series that we haven't shown before. Here the preconception radiation to the mother is the *only* radiation reported. What we find is that the data fits the facts with parameters which are not much different from the previous parameters; in this case the relative risk is 60, not 50, but the overlap of confidence intervals suggests the risks are similar. So we have again confirmed, using a different body of data from the data on the pregnancy exposures, that the radiation is producing serious hazard.

What is particularly useful about this data on mother's preconception exposure is that it rules out some further counterhypotheses which have been raised. The situation here is that, unless you are prepared to believe in sympathetic magic, the only way in which radiation delivered to the mother prior to conception can have an influence on the occurrence of leukemia and other diseases in the child after conception is to have produced genetic damage to the egg cell. This damage is transmitted to the child when the egg is fertilized and later expressed as leukemia. There is really no alternative to a genetic explanation other than outright mysticism or obscurantism. The other nice feature of this purified data is that in effect the leukemic process is started at conception and irrespective of the time when the actual damage occurred. So we have, in effect, synchronized the data here, and this gives some interesting results which I won't have time to go into. However, again, let me emphasize we can clearly rule out the counterhypothesis that the damage is not genetic but some sort of damage to cell membranes, or is transmitted in some other way. It is either genetic or magic.

Now, the next slide [Fig. 13, Slide 7] shows a corresponding data for fathers with the exposure prior to conception. Again there is a good fit, essentially using similar though not identical paramaters. The parameter value is a little lower this time--it's about 25 instead of 50--but that's really not terribly important. What is important is that a model with the notion that radiation is producing both the indicator diseases and leukemia will give an adequate description of what is going on in the real world.

There has been some criticism of our use of 1 percent as a parameter value for the percent affected by the radiation. Whether this parameter value is 1 percent or one-tenth of 1 percent or 2 percent, it does not change the fact that we have shown that serious radiation hazards exist. In practice it is difficult to get a very precise estimate of these numbers from any one body of data. What we must try to do is get more and more information, more and more data, so we can gradually narrow down the estimate. This we now have done using data on combinations of exposures to parents and fetus, and the results are in good agreement with what we previously reported.

WHY THE HEALTH HAZARDS MAY BE EVEN WORSE

Again, let me emphasize, my job is to present a strong prima facie case for hazards. I think that what we actually have here is a much stronger case, much better evidence, and much more conclusive evidence than just a prima facie case. But I am not arguing that this is conclusive because I only need to make a prima facie case for the hazard. This I have done. Now what you have seen so far has been a test of the hypothesis that radiation in the 1-rad range can produce serious health effects. Note that we are talking, in this case, about leukemia in the child and other childhood diseases. These are the diseases we have looked at. This doesn't mean these are the only diseases the genetic damage will produce. I think we are missing a great deal of the health hazard because, for example, the children who have more serious genetic damage as a result of the radiation are not going to make it all the way to age 6 or 7 or 8 and then get leukemia. They are going to die in the first year, in which case they are excluded from this particular survey. Or they may die even earlier, and be counted as the neonatal deaths or stillbirths. Probably, there are also spontaneous abortions produced by genetic damage.

In other words, we are seeing only part of the problem in this particular body of data, perhaps the tip of the iceberg. The genetic damage we can't study is probably producing many more deaths among children than the damage that we can detect. Therefore, we should not permit genetic damage to workers in the nuclear plants by allowing the exposures in the 1-rad range which are currently allowed under the NRC standards. Let me go one step further and say flatly I think that it should be an NRC policy that persons in the childbearing ages, men *or* women, should not be working in areas where there are these exposures.

If somebody has to work there, it should at least be a person who is past the childbearing age, who is not going to put genetic damage into future generations. Because the other side of the coin is that the child may survive despite the damage--this can happen according to our model. That damage will be expressed in the *next* generation or the generation after that. We have put permanent genetic damage into the population. This is something which I think we cannot allow. Therefore, the NRC should set regulations which essentially exclude persons in the childbearing age from this kind of exposure.

Existing regulations to the effect that the woman who has become pregnant should be warned to stay away from radiation (as she might be warned of cigarettes) do not provide protection. The workers, of course as you can see, may be exposed *prior* to pregnancy and still have the damage. Moreover, when women are in early pregnancy they are often not even aware of this. So this kind of regulation is a fraud on the public. We cannot afford to have this kind of paper tiger or fake protection for workers any longer. The Nuclear Regulatory Commission has to bite the bullet and write really protective regulations.

DNA-DEGRADATION: THE UNDERLYING THESIS

And now let me briefly discuss a matter which involves opinion as well as fact. At this point you might wonder: How does the x-ray to the parents actually produce all these diseases in children? I will now say something about processes that lead to leukemia, about what happens between the time that the x-ray does the damage to DNA and the time when leukemia occurs. This necessarily involves some speculations that are not established by factual evidence. We don't know all there is to know about this process. However, we have a fairly good idea of what is going on, and I think it is important to give some explanation, even a somewhat speculative one. There should be no aura of mystery or magic as to why there is a relationship between the exposure to radiation and the occurrence of multiple diseases.

Whenever one talks about underlying mechanism or machinery or processes, there are likely to be lots of technicalities.

However, the main points can be understood without going into these technicalities. For example, we are concerned with the damage produced by the x-ray in the structure of the DNA in human genetic material. This is what I will mean when I speak of "genetic damage." We know that there has to be some kind of genetic damage because the initiation of the process by the x-ray does not immediately produce leukemia. Leukemia takes perhaps 5 to 10 years to develop. Before there can be leukemia, the misinformation that is put into the cell by the x-ray damage has to be duplicated and reduplicated millions of times. We know this because it takes a large white cell population to produce the leukemia. This is not something produced by a single cell.

The initiation of leukemia by the x-ray is a strictly physical process. However, the process that goes on between that point and the time leukemia develops is necessarily a biological process. In fact, it is an evolutionary process. From a mathematical model which I won't discuss today, we could get an idea of how misinformation in a cell can impair feedback machinery that controls the white cell population. It is probably the failure of the feedback control that leads to excessive numbers of white cells. This is what leukemia is all about.

My main point, though, is that the misinformation in the DNA produces, say, the loss of an enzyme. It affects more than that part of the host defense system represented by the white blood cells. Other host defense operations are also impaired or rendered less efficient by this damage. The same breakpoint in the DNA that is responsible for producing the leukemia can be responsible for producing various other diseases as well.

The technical language used in talking about this kind of genetic effect is, I think, very confusing, not just to the public, but to the people who use the language. It's based on outmoded ideas and misconceptions. Unfortunately, the obsolete notions are built right into the technical language and therefore I won't use this language at all.

So let us speak in plain English about the implications of the degradation of DNA by x-ray that I have just described. The first important point is that *it does not take a very great amount of x-ray to produce the kind of genetic damage that will eventually lead to leukemia or other diseases*. All that is involved in initiating the process is disruption of the delicate chemical structure of the DNA, and relatively low levels of energy will suffice to do this.

PROTECTING THE WORKERS AND THE PUBLIC

Let me add a word about higher levels of radiation. Three papers have just appeared in the *Journal of the National Cancer Institute* which show what happens with exposures in the range between 100 rads and 1000 rads. The curves show that up to about 350 rads there is an increasing risk of breast cancer, but at a still higher dose the risk may actually go down. The reason for this is that heavy damage to the DNA produces a cell which is not reproductively competitive. Roughly speaking, you could say that the breast cancer was "caused" and "cured" at the same time.

If you can see this one point, that the limiting factors are different in different dosage ranges, then you can see why we can't expect to learn too much about what happens in the 1-rad range by looking at data on dosages outside this range. Estimates obtained by extrapolating on a straight line assumption from data at high levels of exposure are dubious. The linear assumption breaks down within the high dosage range, so it can hardly be expected to be meaningful in an order of magnitude below this range. *So the only way we can really tell what is going on in the 1-rad range is to look and see what happens to human beings exposed in this range, and that's exactly what we are doing now.*

I would hope that by showing you this factual evidence, all of these tables and graphs and pictures, I have not confused you. What all this boils down to is one simple argument: Suppose we take as a working hypothesis that there is a serious radiation hazard at low levels of ionizing radiation. Suppose that the x-ray produces damage to the DNA that will be expressed in the child later as leukemia and childhood diseases or in the individual himself as leukemia and heart disease. Then we can explain quite adequately and precisely the data actually found in the Tri-State Survey. The hypothesis fits the fact. However, we cannot explain this data without assuming some kind of serious radiation hazard.

There could be some legitimate argument about whether the parameters are exactly right. There can be counterhypotheses about the underlying mechanism that explain how x-ray damage to DNA can eventually produce leukemia and the other diseases. This is really irrelevant. The main point is that without a radiation hazard hypothesis you can't explain this data. With that hypothesis you can.

Note also, this is not data on a couple of hundred rats given a very high dose of saccharin. We are talking about data on about 13 million human beings exposed to diagnostic radiation in the 1-rad range. We are talking about excellent scientific data, really the best of its kind. We are talking about a careful scientific analysis, not just a casual look at the data. We come out with the conclusion that there are very serious human health hazards in the 1-rad range. This is much stronger than a prima facie case, but I am saying that it is *at least a prima facie case for this hazard.*

That means that we have got to stop looking at dosages under 5 rads per year as something we can tolerate. It doesn't matter now whether we are talking about workers in the nuclear industry, or persons who are getting diagnostic x-ray, such as CAT scans. It doesn't matter now exactly what kind of radiation technology is producing the hazard--the point is the hazard is there. We have got to do something about it unless we are willing to give all our children and our children's children a curse that will go on for generations and generations.

Now I think that the data that I have presented represents a prima facie case. If my opponent wants to argue this point, he has to do more than say this or that is wrong. It is up to my opponent to show that there is any other way to explain this data without assuming that there is a very serious radiation hazard in the 1-rad range. He has to present an argument which would justify the claim that there is no serious radiation hazard in this range and that it is safe to leave the level at 5 rems per year for nuclear workers. I don't think he can do this.

chapter 7

WHAT CAN THE PUBLIC LEARN FROM THE NRC MEETING?

How a technical point can be clarified by using several different languages.

INTRODUCTION

After making every effort to present the facts about human health in a comprehensible way, and after answering all of the sensible questions raised about this work and the current picture on low-level radiation hazards, I had resolved to turn to other public health matters. Having spent more than 25 years, off and on, on the cigarette-cancer issue and more than 10 years on radiation hazards, and having seen colleagues become completely identified with a single issue or position, I felt it was time for a change.

However, it is rather harder than it would seem to break loose from these absorbing issues. For one thing, the fact that there is often little or no progress in protecting the public, even over a period of years, makes one feel guilty for not trying just one more time to get effective public action. In the cigarette-cancer issue this "one last effort" did have some effect. It produced a "tar derby" that did help. There is always the hope, however forlorn, that one more paper, one more study, one more push will get results. So, despite my intention to send in a couple of graphs that the NRC had requested and then turn to other matters, I ended up writing a reprise on my presentation.

In pushing for an Adversary Science format with organized critiques as well as cross-examination on both sides, my aim had been to focus the discussion on the basic scientific issues. The hope was to avoid the kind of irrelevant and irresponsible gamesmanship that goes on at unstructured scientific conferences. Hence it was disappointing that the critiques did not focus on the key finding--the doubling dose. The basic questions of data, theory, and technology for rigorous testing were almost untouched. The only point raised against the key findings was plainly wrong. So my intention was to make the technical point as briefly as possible and be done with it.

As the rest of this chapter shows, I changed my mind. True, the point could be made easily in algebraic language, but the trouble was that those who read the transcript but were not mathematically inclined would then have to take my reply on faith. So what I ended up doing was to make the same point in four ways (and in four languages). Whether a reader preferred graphical language, algebraic language, arithmetic language, or plain English, the argument would be there in whatever language was most convincing. This would not, of course, convince anyone in the "radiation protection community" since they wouldn't listen in any language.

However, there might be some concerned citizens who would be willing to proceed in a situation that was a little complicated and where it appeared that the experts disagreed. They might then discover that, in this common impasse in public health issues, it sometimes is possible for the public to tell which expert is right. Perhaps this way the public would learn that it is not necessarily the actual complications of these health issues that tie them up in knots. The public could then see that much of the confusion arises simply because the members of the scientific community fail to meet their responsibilities to the public.

One of the main reasons for the unnecessary "scientific controversies" on the cigarette-cancer issue and a long series of health issues over the past generation has been the intransigence and incompetency of scientists--often on both sides of these issues. The public has had no way of knowing that individuals or institutions are not doing their job or lack the competence to do it. Ordinarily the decision-making process goes on behind the closed doors with, at best token public representation. The public members cannot understand much of what is being said.

While the Adversary Science format used in the NRC meeting can hardly be called a "success" since it failed to get the federal agency to move on the issue, it has at least served one purpose. It gave the public a good opportunity to judge the incompetence of the "radiation protection community" on radiation health issues.

A TEST OF THE EXPERTS

This brief retrospective commentary is primarily addressed to the general public and its elected representatives. Any member of the public who attended the NRC meeting of April 7, 1978, or who reads the transcript of this meeting, is likely to end up rather confused at the end. Many of these persons might automatically suppose that this confusion was due to their own inadequate understanding of the scientific and technical matters discussed. Actually, most of the confusion is a result of the incompetence of some of the professionals who were involved in this hearing. One of the main lessons that the public can learn by careful study of this Washington debut of Adversary Science procedures is this: For too many years the protection of the health and safety of the public against the hazards of low-level ionizing radiation has been entrusted to professionals who are not competent to do this job.

It was disappointing (but not entirely unexpected) that the crucial scientific findings presented at the April 7 meeting were not subjected to cogent, searching criticism and appraisal. However, one unexpected result of the Adversary Science format was to put the public in the somewhat unusual position of being able to judge for themselves the competency of the professionals involved.

Something really new was presented: The first dosage response curve for ionizing radiation in the 1-rad range. Here was the answer to a scientific question that has been hotly debated for years. My formal "adversary"--Dr. Rothman--and my actual adversaries--the self-styled "radiation protection community" which directed so many questions at me--were therefore confronted with a test of *their* professional competence: Do they understand how this dosage response curve was constructed and what it means?

Speaking as a professional biostatistician and epidemiologist with an international reputation to risk, I can state flatly that none of the "adversary" professionals understood this dosage response curve. Any reader who will take the trouble to follow the more detailed explanation that will shortly be given can see for himself or herself that these professionals simply missed the whole point. What makes this particularly instructive is that the point missed was a simple mathematical one, not something requiring extensive biostatistical or epidemiological experience.

A GRAPHICAL ARGUMENT

To put this in plain English, Dr. Rothman made what I would call an "elementary mistake." This isn't the kind of mistake that involves subtle error in judgment. It is the kind of mistake that you want to kick yourself for making after it is pointed out to you. The "radiation protection community" took this mistake as their cue and rushed in to make the same mistake over and over again. This is hardly the mark of critical thinking, independent judgment, or professional competence.

At issue here is the question: Is the y-axis of the dosage response curve a measure of the biological effects of radiation? In one form or another, this question was asked all afternoon. As I pointed out in my rebuttal (but to no avail), there is a simple mathematical relationship between the response variable that I had used ("percent affected") and a response variable directly related to the risk of leukemia. Indeed, in his discussion of the childhood leukemia Dr. Rothman had tacitly employed as his measure of response to radiation a quantity that could be called the "percent excess risk of leukemia." For practical purposes, there was a very easy way to go from the much-questioned variable to one where there was no reasonable question. *All that was needed was to relabel the y-axis of the graph.* That's all.

In response to a request from Dr. Kastner of the NRC, and to illustrate this point in a way that any reader can follow, a new set of dosage response curves are presented where the y-axis has been relabeled. Figure 14 shows the dosage response curve and confidence intervals for the men over 65 and corresponds to Slide 11 [Fig. 2].

Figure 15 gives the dosage response curves for the three separate age groups and corresponds to Slide 13 [Fig. 3]. *Anyone can verify by direct comparison of these graphs that the only difference is that the y-axis has been relabeled and is now the "percentage excess leukemia risk."*

The vexing question of the y-axis was about the only specific objection that was raised to this crucial dosage response curve for the men exposed to ordinary diagnostic radiation (although this should have been the focus of a critique of this work). Once this information is accepted as scientifically valid, there are immediate scientific and public health implications that follow from these results. A direct statistical consequence is that the estimate of the doubling dose for leukemia is in the 1-rad range, probably less than the 5 rems permitted annually by the NRC.

No other federal agency permits exposures to known human carcinogens at the doubling dose. Ordinarily, the permissible levels are less than 1 percent of such a dose. Only for this one carcinogen (ionizing radiation) has the setting of protective standards been controlled by persons with a vested interest in the technology. What the new findings mean, to put it bluntly, is that the "radiation protection community" has been getting away with murder for too many years.

AN ALGEBRAIC ARGUMENT

Sometimes saying that same thing in a different way helps to make a point. A graphical answer to the question of the y-axis has been given, but a more precise way to make this point is in simple algebraic terms. This line of argument should make the point crystal clear to anyone such as the members of the "radiation protection community" from a "hard" science. The formulas are so simple here that high school algebra is all that is needed. Dr. Rothman himself describes the formula *in words* when he makes the following comments on the childhood leukemia data:

> Until we have such knowledge of predictors, however, we consider all individuals as equally at risk and as sharing the overall risk. So all persons exposed to radiation share an estimated 50 percent increase in risk. What if we pretend to be able to narrow down, or refine, the susceptible group? For example, suppose the extra risk could be concentrated in half of those exposed to radiation. Then, instead of a 50 percent increase shared by everyone, half of those exposed would have a 100 percent increase, that is a doubling risk. Suppose the extra risk could be concentrated in ten percent of exposed individuals. These ten percent would have a five-fold increase in risk. The smaller the size of the group that one supposed to be susceptible, the greater the increase in risk would have to be in that group to be consistent with an overall relative risk for all exposed individuals of 1.5. Dr. Bross has supposed a susceptible group which is one percent of all individuals; therefore, the relative risk in this hypothetical one percent must be 50 to give an overall relative risk of 1.5. If one considers a smaller size for the susceptible group, say one-hundredth of one percent, the relative risk in the susceptible group will have to be 5000.

(Statement is on pages 9 and 10 of the Rothman critique.)

Now what is Dr. Rothman saying here? He has given numerical examples of a simple formula. In algebraic symbols let:

A = proportion affected by radiation exposure

R_L = relative risk of leukemia in those affected

PI = percentage increase in risk in those affected

PE = percentage excess leukemia risk

Then Dr. Rothman was giving examples of the formula

$$PE = A \times PI$$

The "percentage increase" is

$$PI = 100(R_L - 1)$$

so that

$$PE = A \times PI = A \times 100(R_L - 1) = K \times A$$

where $K = 100(R_L - 1)$. In other words, the "percentage excess leukemia risk" is simply proportional to the "proportion affected." Moreover,

$$PE = 100\ AR_L - 100A$$

where AR_L was called the "product term." If R_L is large (e.g., of the order of magnitude of 10), then the product term largely determines PE, and the results will not depend in any critical way on the separation of the "product term" into its components.

What I have just presented is not of overwhelming mathematical difficulty either to statisticians or to any concerned citizen who knows elementary algebra. Nevertheless, Dr. Rothman on page 12 of his critique says:

> The vertical axis in this graph is not cancer risk, as it should be. Notice it is labeled percent affected, with affected in quotation marks. The quotation marks signify, I think, that the affected group is not affected at all. We know for sure what it does signify, but I suppose it is the size of the group that Dr. Bross never explained how these estimates of the size

of the susceptible group were derived, but in any case, the curve is *not* a dose response curve of leukemia risk.

What Dr. Rothman is saying is that while the "percentage excess leukemia risk" is a perfectly sensible measure to use for reporting the risk of radiation exposures, if one divides by a constant, approximately by a factor of 10, then the curve is "not a dosage response curve of leukemia risk. The algebraic language brings out clearly a point which might be confused in a verbal language.

The connection between the algebraic equations and the graphs may be clearer with a little arithmetic.

AN ARTHIMETIC ARGUMENT

To make slide 11 [Fig. 2] more meaningful to some of our readers, we have converted the vertical axis from "percent affected" to "percent excess leukemia risk." To do this the value of each point on the vertical axis has been multiplied by 9; for example, the point 5 on the "percent affected" graph is now 45, and the point 20 becomes 180 on the "percent excess leukemia risk" scale. For a given trunk radiation dose, on the new graph we can now read the *excess* leukemia risk percent; for example, for 13 rads, the *excess* leukemia is 200 percent. Since the risk of leukemia has been increased 10-fold ($R_L = 10$), the *excess* leukemia risk is (10 - 1) or 9.

The same conversion has been made for slide 13 [Fig. 3] that presents similar data for the three age groups.

(At this point the NRC text goes through the same argument made in chapter 6. It is interesting that even though this error was never acknowledged, it did not produce the long-lasting confusion that such errors usually produce. Subsequent hostile critiques did not repeat this mistake, so apparently the point did get across in the algebraic language.)

Any intelligent reader can go over the algebraic formulas that have just been given and see for himself or herself that there is a very simple relationship between the measure of response that Dr. Rothman was talking about for children, the "percent excess leukemia risk," and the response variable "percent affected" in the dosage response curve. There is another point that can be seen at the same time: *The findings from the dosage response curve for adults does not depend upon (or require) the separation of the product term into its components.* This is because in the "percent excess leukemia risk" we are essentially dealing with the product term rather than the separate components (e.g., the "percent affected").

Therefore, the technical point raised by Dr. Charles Land of the Environmental Epidemiology Branch at the National Cancer Institute does not really have to come into the argument (although it can also be answered directly as was done in my rebuttal statement). What a careful reading of this transcript will, therefore, show is that no pertinent objections have been raised to the dosage response curve, the doubling dose estimates that derive from it, or to the immediate implication that the 5-rem level poses serious hazards and should no longer be permitted by the Nuclear Regulatory Commission.

AN ARGUMENT IN ENGLISH

Some thoughtful readers might wonder why I used "percent affected" when I could have avoided the whole argument by using "percent excess leukemia risk." I had, in fact, considered doing this. However, the "percent affected" provides a useful answer to the argument which always comes up: Leukemia is such a rare disease that we need not worry about it. The answer to this is that leukemia is our most sensitive early warning signal of other cancers and other diseases. The "percent affected" gives a quantitative measure of the extent of serious genetic damage that can lead to leukemia *and* also to these other diseases.

For example, the Bureau of Radiological Health estimates that of the more than 200 million medical and dental x-rays given each year, 30 to 40 percent are unnecessary. I would estimate at least 10 million unnecessary medical x-rays. What is the magnitude of the public health problem produced by these unnecessary x-rays? If we look at excess leukemia alone, we cannot see the real magnitude of the problem. However, among the persons receiving the unnecessary x-ray, at least 5 percent are going to be "affected" by this x-ray in the sense of sustaining serious genetic damage. Clearly, we are now talking about tens of thousands of persons, not just a handful of people.

Perhaps I should be grateful to Dr. Rothman for going off on this false scent and leading the whole pack of engineers and physicists from the "radiation protection community" away from any authentic critique of our findings. He has thus provided the public with an easy way to evaluate the competence of these persons. It is a bit ludicrous that those experts who prepare the complicated environmental impact statements on health hazards for the NRC are not capable of seeing the very simple algebraic relationships involved in our work.

So perhaps with the help of the Adversary Science format, the public can see that our facts stand, our analysis has not been faulted, and our findings have not been seriously challenged. From the dosage response curve, the doubling dose has been estimated, and it has been shown that the hazards of low-level ionizing radiation are about 10 times worse than previously estimated by linear extrapolation in the BEIR Report.

FIGURES

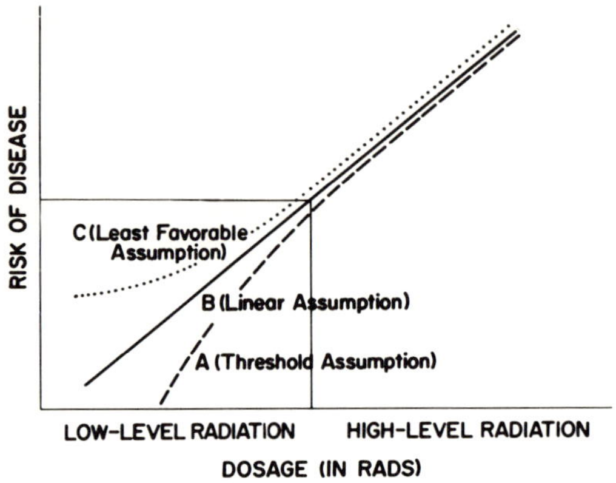

Fig. 1 (Slide 1) Three hypotheses about the relationship between risk of disease and exposures to low doses of ionizing radiation (in rads). From NRC presentation of April 7, 1978.

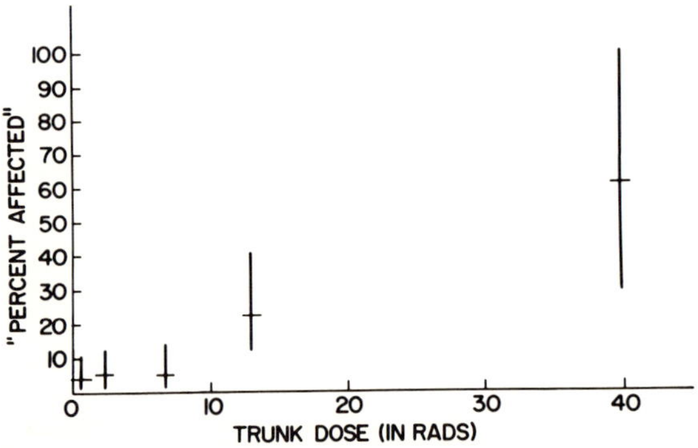

Fig. 2 (Slide 11) Percent affected by trunk dose of x-ray (in rads) for men over 64 years of age. From the Amer. J. Pub. Health, Vol. 69, No. 2, 1979, with permission.

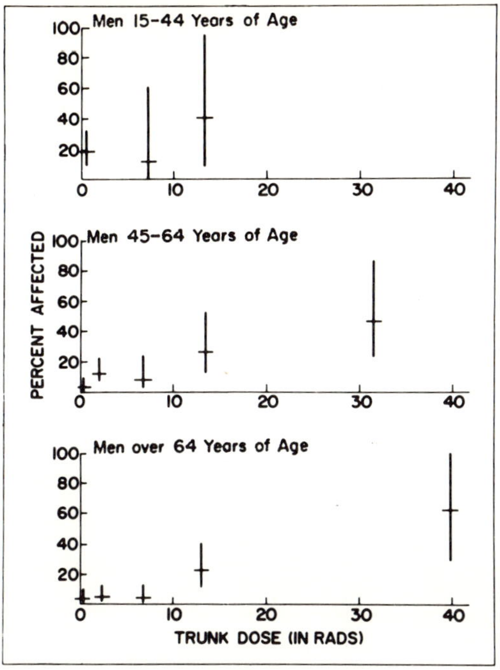

Fig. 3 (Slide 13) Percent affected by trunk dose of x-ray (in rads) for men in three age groups. From the Amer. J. of Pub. Health, Vol. 69, No. 2, 1979, with permission.

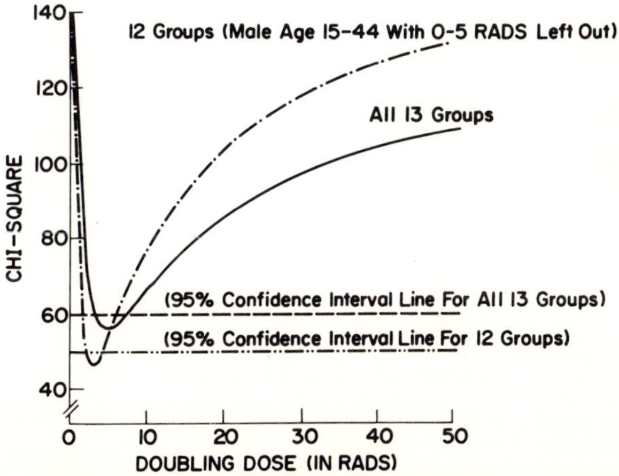

Fig. 4 (Slide 14) Calculated chi-square for doubling doses 0-50 rads including and excluding an outlier. From NRC presentation of April 7, 1978.

	65 Years	
	No heart disease	Heart disease
Less than 1 rad		
Leukemia		
Observed	14	9
Expected	17.43	8.27
No leukemia		
Observed	17	5
Expected	17.92	4.98
1-5 rads		
Leukemia		
Observed	19	9
Expected	17.43	9.35
No leukemia		
Observed	17	4
Expected	16.98	4.88
5-10 rads		
Leukemia		
Observed	9	7
Expected	12.38	6.56
No leukemia		
Observed	10	5
Expected	12.14	3.47
10-20 rads		
Leukemia		
Observed	13	10
Expected	11.68	11.76
No leukemia		
Observed	4	4
Expected	6.62	2.74
20 rads or more		
Leukemia		
Observed	6	5
Expected	4.66	6.40
No leukemia		
Observed	1	1
Expected	1.15	0.93

Fig. 5 (Slide 8) Observed and expected number of men over 64 by leukemia and heart disease status and calculated radiation exposure (in rads). From the Amer. J. Pub. Health, Vol. 69, No. 2, 1979, with permission.

45-64 Years

	No heart disease	Heart disease
Less than 1 rad		
Leukemia		
Observed	12	5
Expected	15.98	3.72
No leukemia		
Observed	49	5
Expected	48.89	7.47
1-5 rads		
Leukemia		
Observed	14	10
Expected	18.60	7.13
No leukemia		
Observed	40	1
Expected	36.01	6.78
5-10 rads		
Leukemia		
Observed	5	4
Expected	9.81	3.33
No leukemia		
Observed	18	7
Expected	22.22	3.87
10-20 rads		
Leukemia		
Observed	13	1
Expected	10.58	4.99
No leukemia		
Observed	12	4
Expected	13.48	3.21
20 rads or more		
Leukemia		
Observed	5	5
Expected	6.59	3.49
No leukemia		
Observed	7	0
Expected	5.56	1.74

15-44 Years

	No heart disease	Heart disease
Less than 5 rads		
Leukemia		
Observed	26	1
Expected	25.01	1.95
No leukemia		
Observed	47	2
Expected	48.97	2.12
5-10 rads		
Leukemia		
Observed	2	0
Expected	2.59	0.17
No leukemia		
Observed	6	1
Expected	7.03	0.27
10 rads or more		
Leukemia		
Observed	1	1
Expected	4.49	0.39
No leukemia		
Observed	5	1
Expected	5.94	0.31

Fig. 6 (Slide 10) Observed and expected number of men over 64 by leukemia and heart disease status and calculated radiation exposure (in rads). From the Amer. J. of Pub. Health, Vol. 69, No. 2, 1979, with permission.

Rads = 0.167 (no. of chest x-rays) + 0.790 (no. of abdomenal x-rays)

Numerical values came from Public Health Service X-Ray Exposure Study: Population Exposure to X-rays U.S., U.S. Department of Health, Education and Welfare, 1964, 1970.

Fig. 7 (Slide 9) Formula for calculation of x-ray doses. From NRC presentation of April 7, 1978.

Case 1: No radiation exposure reported

The notation used is as follows:

X_1 = age group (0 = 1-4 yr; 1 = 5-9 yr; 2 = 10-14 yr)

X_2 = occurrence of indicator disease (0 = no, 1 = yes)

X_3 = occurrence of leukemia (0 = no, 1 = yes)

$D_0(X_1)$ = probability of at least one indicator disease

$L_0(X_1)$ = probability of leukemia

K = no. of yr in survey

The probability of a given status with respect to indicator disease, leukemia, and age is

$$P_0(X_1,X_2,X_3) = [D_0(X_1)]^{X_2}[1 - D_0(X_1)]^{1-X_2}[L_0(X_1) \times K]^{X_3} [1 - L_0(X_1) \times K]^{1-X_3}$$

The expected numbers in a 2 × 2 table can be calculated using the following:

$N(X_1)$ = no. of children in population in age group X_1

$H_0(X_1)$ = proportion with no radiation exposure

S = sampling rate in random sample

So that $N(X_1) \times H_0(X_1)$ is the total number with no radiation exposure reported in the population in the age group X_1, and then the expected numbers for the individual cells in the 2 × 2 table are

$$E_0(X_1,X_2,X_3) = N(X_1) \times H_0(X_1)S^{1-X_3} \times P_0(X_1,X_2,X_3)$$

(Note that when there is leukemia, the sampling rate is 1.)

Fig. 9A (Slide 3) Mathematical model for occurrence of leukemia and indicator diseases in children when no radiation exposure is reported. From JAMA, Vol. 237, No. 22, 1977, with permission.

	Indicator disease (no.)	
	No	Yes
Leukemia		
Observed	27	3
Expected	27.7	2.9
No leukemia		
Observed	48	5
Expected	47.3	4.9

Fig. 8 (Slide 2) Observed and expected number of children aged 1-4 by leukemia and indicator disease status in the absence of reported radiation exposure. From JAMA, Vol. 237, No. 22, 1977, with permission.

Case 2: X-ray during pregnancy reported

The notations used here are as follows:

D_1 = probability of indicator disease in group exposed to radiation and affected (i.e., suffered genetic damage)

L_1 = probability of leukemia in affected group

A = proportion affected by radiation exposure

H_1 = proportion with radiation exposure

So that $D_1 = D_0 \times R_D$ and $L_1 = L_0 \times R_L$, where

R_D = relative risk of indicator disease

R_L = relative risk of leukemia

As before, $N(X_1) \times H_1(X_1)$ is the total number with reported x-ray during pregnancy in the age group X_1 in the population.

Now, the probabilities for individual cells in the 2 × 2 table with radiation exposure contain two parts: The probabilities for (1) those who were exposed to radiation but not affected, and (2) those who were affected by radiation. The probabilities for the former are the same as those with no radiation exposure reported, and the probabilities for the latter are

$$P_1(X_1,X_2,X_3) =$$

$$[D_1(X_1)]^{X_2}[1 - D_1(X_1)]^{1-X_2}[L_1(X_1)K]^{X_3}[1 - L_1(X_1) \times K]^{1-X_3}$$

So, the expected values for the group with exposure to mother's radiation during pregnancy are

$$E_1(X_1,X_2,X_3) =$$

$$N(X_1)H_1(X_1)S^{1-X_3}[(1 - A)P_0(X_1,X_2,X_3) + A\,P_1(X_1,X_2,X_3)]$$

Fig. 9B (Slide 3 continued) Mathematical model for occurrence of leukemia and indicator diseases in children when x-ray during pregnancy is reported. From JAMA, Vol. 237, No. 22, 1977, with permission.

	Model Indicator disease (no.)		Adjusted model Indicator disease (no.)	
	No	Yes	No	Yes
Leukemia				
Observed	45	12	45	12
Expected	32.7	3.4	42.3	12.0
No Leukemia				
Observed	60	3	60	3
Expected	55.7	5.8	56.0	6.1

Fig. 10 (Slide 4) Observed and expected number of children aged 1-4 by leukemia and indicator disease status when x-ray during pregnancy is reported. From JAMA, Vol. 237, No. 22, 1977, with permission.

	Indicator diseases (no.)			
	No radiation		Radiation	
Age group (yr)	No	Yes	No	Yes
1-4				
Leukemia				
Observed	27	3	45	12
Expected	27.7	2.9	42.3	12.0
No leukemia				
Observed	48	5	60	3
Expected	47.3	4.9	56.0	6.1
5-9				
Leukemia				
Observed	13	3	11	7
Expected	18.2	3.0	14.0	6.9
No leukemia				
Observed	79	13	53	8
Expected	77.8	12.8	51.2	8.8
10-14				
Leukemia				
Observed	8	1	4	4
Expected	11.9	2.6	5.1	3.8
No leukemia				
Observed	72	16	30	6
Expected	68.8	15.3	27.9	6.5
Total				
Leukemia				
Observed	48	7	60	23
Expected	57.8	8.5	61.4	22.7
No leukemia				
Observed	199	34	143	17
Expected	193.9	33.0	135.2	21.4

Fig. 11 (Slide 5) Observed and expected number of children by age, by leukemia and indicator disease, and by report of exposure to x-ray during pregnancy. From JAMA, Vol. 237, No. 22, 1977, with permission.

	Indicator diseases (no.)			
	No radiation		Radiation	
Age group (yr)	No	Yes	No	Yes
1-4				
Leukemia				
Observed	27	3	19	7
Expected	27.9	2.9	24.0	7.3
No leukemia				
Observed	48	5	30	4
Expected	47.6	4.9	20.5	3.3
5-9				
Leukemia				
Observed	13	3	4	4
Expected	18.5	3.0	4.5	2.5
No leukemia				
Observed	79	13	19	0
Expected	78.7	12.9	16.2	2.8
10-14				
Leukemia				
Observed	8	1	0	1
Expected	11.9	2.7	1.3	1.1
No leukemia				
Observed	72	16	8	1
Expected	68.3	15.6	6.9	1.7
Total				
Leukemia				
Observed	48	7	23	12
Expected	58.3	8.6	29.8	10.9
No leukemia				
Observed	199	34	57	5
Expected	194.6	33.4	53.6	7.8

Fig. 12 (Slide 6) Observed and expected number of children by age, by leukemia and indicator disease, and by report of exposure of mother to x-ray prior to conception. From NRC presentation of April 7, 1978.

	Indicator diseases (no.)			
	No radiation		Radiation	
Age group (yr)	No	Yes	No	Yes
1-4				
Leukemia				
Observed	27	3	36	13
Expected	27.7	2.9	38.9	14.9
No leukemia				
Observed	48	5	65	9
Expected	47.3	4.9	66.4	7.6
5-9				
Leukemia				
Observed	13	3	9	6
Expected	18.2	3.0	10.9	5.0
No leukemia				
Observed	79	13	43	12
Expected	77.8	12.8	46.6	8.2
10-14				
Leukemia				
Observed	8	1	4	3
Expected	11.9	2.6	3.5	1.9
No leukemia				
Observed	72	16	19	7
Expected	68.8	15.3	20.9	5.0
Total				
Leukemia				
Observed	48	7	49	22
Expected	57.8	8.5	53.3	21.8
No leukemia				
Observed	199	34	127	28
Expected	193.9	33.0	133.9	20.8

Fig. 13 (Slide 7) Observed and expected number of children by age, by leukemia and indicator disease, and by report of exposure of father to x-ray prior to conception. From NRC presentation of April 7, 1978.

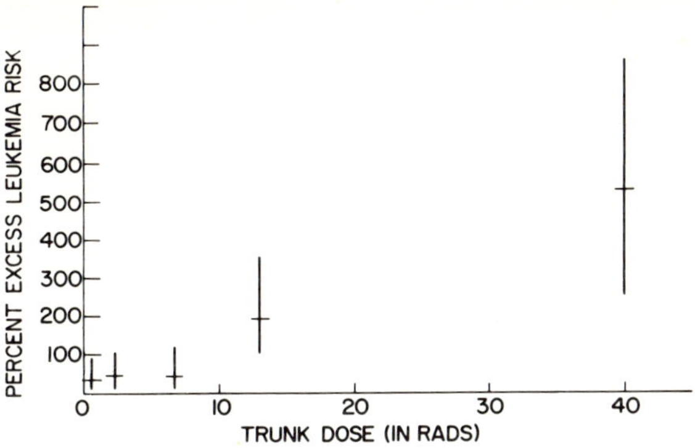

Fig. 14 Percent excess leukemia risk by trunk dose of x-ray (in rads). Compare to Fig. 2.

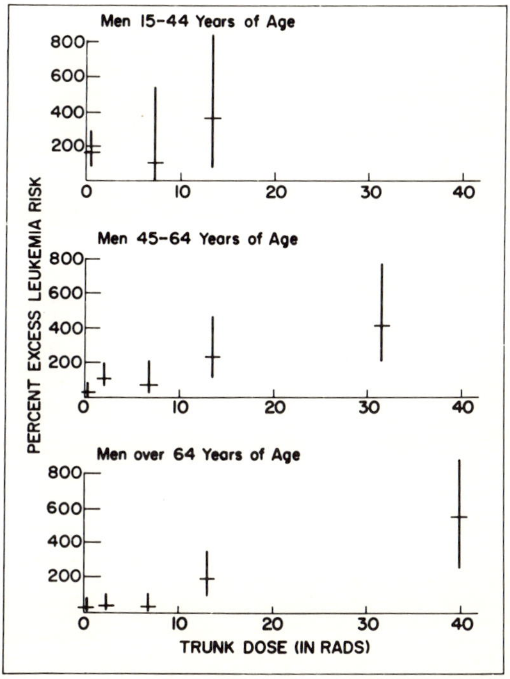

Fig. 15 (Percent excess leukemia risk by trunk dose of x-ray (in rads). Compare to Fig. 3.

chapter 8

JUDGING A SCIENTIFIC ARGUMENT

Criteria for judging quality of data and the adequacy of theory in the evaluation of scientific validity.

GETTING TO THE TRUTH

Nowadays much of our information comes through the mass media, where it is mixed in with a great deal of misinformation, nonsense, trivia, and drivel. Scary reports of new hazards are followed by bland denials that there is any hazard. Perhaps somewhere in this mixture is the one piece of information that might save your life--but how would you recognize it? How can you get to the truth?

As we've already seen, there are scientific strategies such as the Galilean rule that enable a scientist to draw valid conclusions from factual evidence. What I will try to do in this chapter is to give you a clearer picture of how this is done. My NRC presentation will be used as a model for the process. It can serve as a theoretical model of the general procedure and also as a practical model of how the strategies can be applied to a particular body of facts. Perhaps if you can see how a genuine scientific study works, you will be less likely to be fooled or misled by what is presented in the mass media in the name of science. The discussion will be a little more abstract and somewhat more technical than the previous chapters, so you might want to scan the sections without going into the details (or perhaps even go on to Chapter 11).

UNDERSTANDING A SCIENTIFIC ARGUMENT

To better understand the complexities of a scientific argument, it is helpful to break it down into simpler and more comprehensible components. This kind of analysis of a complex process can be shown graphically by means of "block diagrams" such as the one below. For example, a scientific argument is the formal process that we use to go from basic data to reliable conclusions. This concept can be represented by using blocks for the components:

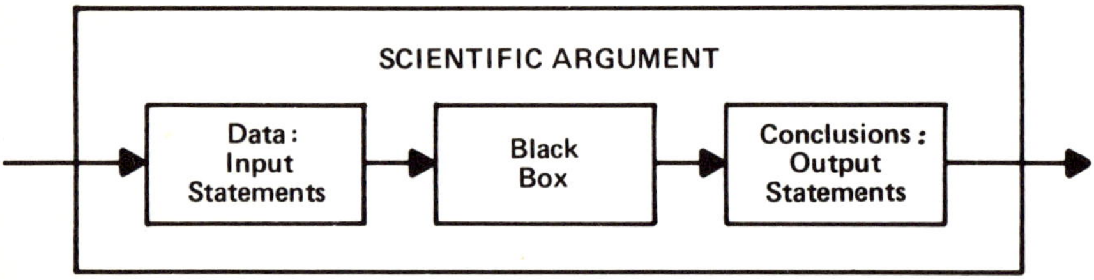

Scientific argument, black box.

Note that the middle box has been labeled "black box." This terminology suggests that, while there is some process that will enable us to go from data to conclusions, we are not yet ready to say anything more about it. One advantage of a black-box analysis is that we can talk about a process without giving the misleading impression that we know all about it. If we label a box with some positive-sounding term (e.g., "scientific method"), we can delude ourselves or others about how much we really know.

There are several advantages in this graphic representation. One is that we can immediately "see" some features of the structure. For instance, we can see that an ordinary argument that makes a general point from specific examples has a similar structure. We can also see that the boxes can be relabeled if we want to be more specific. For instance, the scientific argument in the NRC presentation might be relabeled:

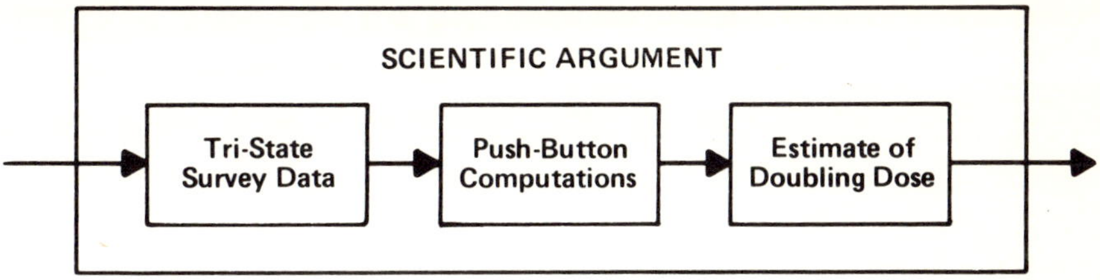

Scientific argument, actual.

This schematic suggests that the computerized biostatistical technology that might be mind-boggling for some concerned citizens can be considered as a "black box." Looking at this point in the opposite way, the abstract "black box" has a concrete embodiment in the push-button operations.

The black-box analysis can be continued by taking the lid off the box and showing what lies inside. Very often this consists of components that are also black boxes. This Chinese puzzle-box structure is typical of complex processes. We may gain a better understanding of what is going on without necessarily achieving a complete understanding.

BLACK-BOX OPERATIONS

If we simply consider the input and the output, we can often say something about what goes on in a black box without actually looking inside. For example, the input is a set of specific statements. The output is generally a much smaller set of more general statements. So we can see that the operations that go on inside the black box are operations on statements. In other words, the processes in the black box are the processes of language. While not necessarily the processes of ordinary or natural language, they must nevertheless involve *linguistic* operations.

What do these linguistic operations accomplish? One thing they obviously accomplish is what can be called "compression." The Tri-

State Survey data feeding into the black box represents an enormous number of statements. While not all of the statements would come into a given analysis of the data, the original data base involves an interview schedule with more than 100 questions administered in approximately 5000 households. So the number of statements going into the black box is of the order of magnitude of hundreds of thousands.

A relatively small number of statements will be coming out of the black box. There may be a single key finding from a study. Few scientific abstracts contain as many as half a dozen findings. So we can say that the operations in the black box somehow compress the large number of statements going into the box down to a relatively few statements coming out of it.

What has this compression accomplished? What is gained by boiling down many statements into a few? For one thing, as human beings we have trouble understanding, communicating, or remembering large numbers of statements. A few statements may be much more manageable. We often complain because oversimplified, general statements can be misleading, but in practice we constantly resort to this kind of compression. We do this in everyday discourse, so it shouldn't be surprising that a similar process occurs in scientific discourse.

We also have to ask: What has been lost by this compression? Clearly, in going from a large number of statements to a small number, something has been lost. But what? Has essential information been lost? Has contact with reality been lost so that the output statements are no longer informative? There are real dangers in any process of compression.

REPRESENTING THE GALILEAN RULE

Chapter 5 has emphasized the role of the Galilean rule in a scientific argument, and we can reemphasize that by taking the lid off the first black box to reveal three more black boxes inside as indicated below:

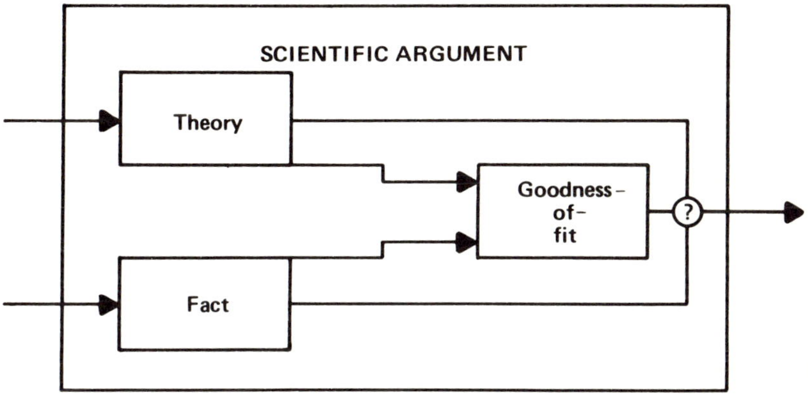

Scientific argument, Galilean rule.

The above schematic illustrates a further use of block diagrams to show the flow pattern of the process. Here, for instance, the output from the goodness-of-fit test is shown with three possibilities. With a satisfactory fit, the output may be a definite conclusion. However, with a poor fit, we may have to modify the theory (shown by the arrow going back to the "theory" box). If the results are inconclusive, we may need more data.

The abstract components of the above schematic can be given a concrete realization by going back to the NRC presentation. The "theory" here is essentially what appears in Figs. 9A and 9B. Figure 9A shows the theory for childhood leukemia, but there are only minor modifications for the theory used for adults [Fig. 9B]. The "data" consists of the observations in Figs. 5 and 6. The goodness-of-fit component is not as easily illustrated, but the general idea is conveyed by Fig. 4. The customary statistical measure of goodness-of-fit, the total chi-square, is shown on the vertical or y-axis of Fig. 4. A range of values for the doubling dose is shown on the x-axis.

Since the smallest value of chi-square corresponds to the value of the doubling dose where there is the best fit between the theory and the facts, it is plausible to take as an estimate of the doubling dose the minimum chi-square estimate. As can be seen from Fig. 4, this leads to the conclusion that the doubling dose for leukemia is probably less than the permissible NRC dosage of 5 rads. The figure illustrates the goodness-of-fit procedure. However, in actual operations this search for the minimum was automatic, and the computer output gave the numerical estimate of the minimum value.

This association of the three figures of the NRC presentation with the blocks in the schematic of a scientific argument somewhat oversimplifies the process, but it may help to make the abstract components more meaningful. We will come back to some of the complications in later sections.

JUDGING DATA QUALITY

In ordinary discourse, the credibility of incoming reports is judged by experience--whether the person has made reliable statements in the past, whether it agrees with reports of other witnesses, and whether it fits patterns of experience with similar events. The informal machinery for keeping false or misleading reports of events out of ordinary discourse includes social pressures against those who don't tell the truth, legal penalties for perjury or slander, and tacit exclusion of statements by those persons who have been unreliable in the past. The processes are similar, but more formal, in scientific discourse.

In science, however, the emphasis shifts from "personalities" to "methodologies" (although this is sometimes only a name change, and personalities are still very much involved). From the present standpoint, the "methodology" of a study can simply be regarded as the machinery for generating the basic reports and carrying through the scientific argument. One question is: Does this machinery generate reliable reports? If the machinery is carefully set up and operated (i.e., if the methodology is sound), then the statements produced would be "admissible." In other words, they would be of high enough quality to allow into the black box.

The Tri-State Survey data had a number of important design features which insured the quality of the incoming data and which thereby supported its admissibility. These features include population-based registries, concurrent random samples of the general population, double-blind interviewing, careful validation of the radiation exposures against hospital and other records, and a generally high level of competence in setting up and carrying out this survey.

Many of these features were 20 years ahead of the times and are lacking in all but the very best epidemiological studies that are carried out today. At the time of writing, there is probably no comparable body of data on diagnostic radiation available for scientific analysis. Few epidemiological studies that have ever been done come up to these high standards. So one might say that if any epidemiological data is admissible to the black box, the Tri-State Survey data certainly is.

In any human enterprise, including science, using good technology may not be enough. It can make a difference just how the machinery is actually operated--and this depends on *who* is running the machinery. In a survey, the double-blind interview technique tends to avoid interviewer biases *if* it is properly done. As it happened, the man who conceived and guided the Tri-State Survey, Dr. Abraham Lilienfeld, has an impeccable reputation and has built this reputation on a long series of outstanding studies. Also participating in the survey were Dr. Morton Levin, Dr. Leonard Schuman, Dr. Saxon Graham, and other reputable investigators. I was involved in the sampling plans. Although science is supposed to be objective, and although rating individual reputations is highly subjective, past experience with the reliability of studies comes into admissibility of data in practice. Good scientists respect good data and the scientists who produce good data.

A schematic for the data is shown below. It emphasizes quality control of the fact-finding process. The feedback arrows from quality control show how evaluation of the incoming data may lead to some revision in the original study design or changes in the data-collection operation. In short-term studies, the defects picked up in the quality-control step may result in doing a whole new experiment:

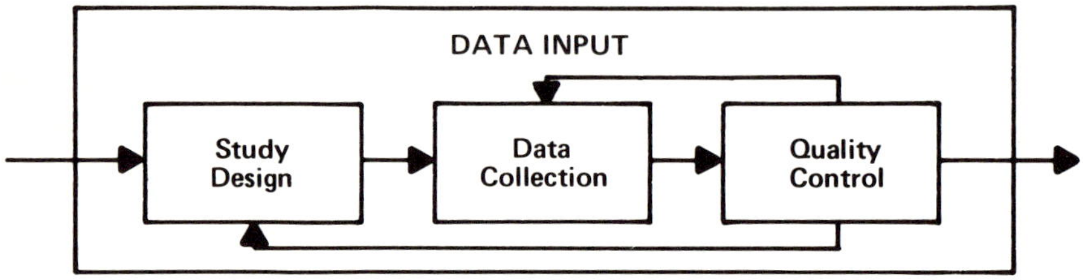

Data input, general.

The overall process is, therefore, called "evolutionary" in the sense that the original plans and procedures may be modified or improved as time goes by. This evolutionary improvement is characteristic of scientific disciplines that are alive and well. *Notice how quality control and critical evaluation are essential if evolutionary improvement is to take place.*

CONTROL OF BIAS

Collection of reliable statements as input takes a great deal of hard work, but it also takes clear thinking and good planning. Care taken in setting up the machinery for collecting data, sound methodology, is what good scientists look for in judging a study. The double-blind interview technique is a good example of the kind of machinery needed to insure high-quality, admissible data.

The purpose here is to avoid biases that might arise if persons being interviewed or the person doing the interviewing had a conscious or unconscious wish to influence the results. For instance, suppose an interviewer had strong opinions that radiation was dangerous (the survey was taken at a time when this view was not widely held, but possibly this view might have been held by some of the participants).

If so, the interviewer might probe more deeply into radiation exposures of the households with a leukemic case member than those of the households in the random sample.

To control this bias, the interviewing was done "blind." The interviewer was not told which kind of household was being interviewed, and the questions that might reveal this were kept to the end of the schedule. The person interviewed was told this was a health survey and was not informed of the real purpose of the survey. This was, of course, long before the days of informed consent. Recognizing that by the end of a lengthy question schedule a trained interviewer might guess the status of the household, the sensitive questions were put earlier in the interview. Simple common sense goes a long way in science as in everyday life.

The machinery for double-blind interviewing is quite simple (though it adds considerably to the cost of the survey). Relatively few epidemiological studies include these and the other methodological refinements of the Tri-State Survey. This is largely because most such studies are done by clinicians who know little about methodology and care even less.

In contrast to the careful design and elaborate precautions of the Tri-State Survey, most biomedical data are collected in such a casual and careless way that they are of doubtful admissibility. Unsubstantiated and even fictional anecdotal material comes into clinical discussions about as often as it comes into everyday discourse.

Unless quality control is stringent for the data going into the black box, the statements coming out of the box will be based on misinformation and will not be reliable either for scientific or policy-making purposes. The conditions for admissibility in the control of the input tend to be codifications of scientific experience. Thus, double-blind interview methods reflect past experience with distortion of results from interviewer biases. Hence, the conditions for admissibility of factual evidence in scientific discourse are going to be peculiar to the scientific enterprise (law and other enterprises set very different conditions) and are fairly specific to the research area.

For instance, a biochemist has to worry about chemical artifacts rather than interviewer biases. Therefore, his methodology will be different from an epidemiologist's. But despite great differences in the construction of statement-generating machinery in the different sciences, the object is still quality control.

The criteria for what is admissible evidence will depend on the methodological state of the art in a given research area, particularly as reflected in the Methods and Procedures sections of published papers. The technical details will be specific to the science, but the general ideas behind these details of quality control are similar to the guidelines for accuracy and honesty in ordinary discourse.

Ironically, the valiant efforts of good scientists to avoid biases in their research often are frustrated by the very subjective way that decisions on presentation, publication, and reporting are made. These put the biases (particularly bias toward the prevailing opinions) back into the literature of the disciplines.

JUDGING DATA INPUT

In the previous section we have been considering the criteria for admissibility of evidence and the question: What information about the world do we let into the black box? Now we turn to the question: What statistical or other operations on this information can be used to insure that the data going into the black box will be trustworthy? From a linguistic standpoint, all of the operations we carry out on the data from here on are purely linguistic--they are the operations in natural languages that linguists called "transformations." They can be called "information-free" or "informationless" because they cannot add any further real-world information. However, they can (and sometime do) lose some of the information.

The schematic below may give some idea of what we would see if we took a closer look at the operations in the data input:

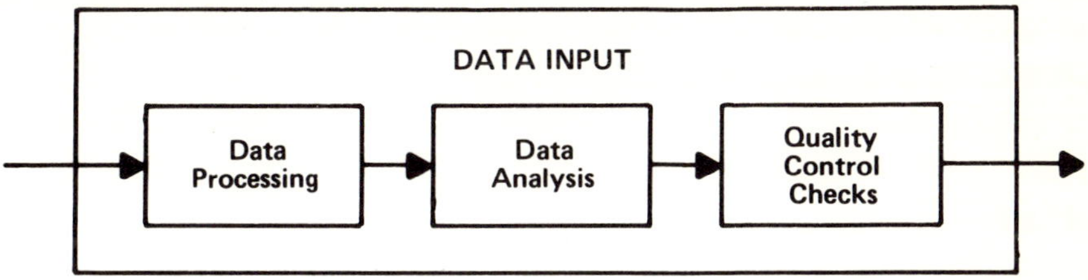

Data input, specific.

The operations in data processing are examples of linguistic transformations. Some are reversible and do not lose information. For instance, if the age of an individual is reported, this number may be carried through all the steps in data processing (assuming no errors are made). It may, therefore, be written on a paper "code sheet," punched into punch cards, or transformed to a magnetic character on tape. In principle, it could always be printed out again as an ordinary number. Such processing operations are called "reversible," and no information is lost.

Other "code sheet" operations lose information. However, the information that is retained is in a form which facilitates the subsequent statistical analysis. For instance, the record may mention a specific x-ray procedure. However, the code sheet may simply record it as x-ray to the trunk. This item of information may end up on a computer-readable tape, but there is now no way to reverse the process. It is not possible to recover the information about the specific x-ray procedure. It has been lost. This is one example of an irreversible linguistic transformation.

The loss of information may or may not affect the validity of a study. In general, if there are positive findings, loss of information from using broad overall classes instead of narrower subclasses (unless there is a bias in the process) would not impair the validity of the positive finding. On the other hand, if the findings are negative, then the loss of information might raise serious questions. What few scientists realize is that establishing negative conclusions is really much harder (in a technical sense) than establishing positive ones. Admissibility questions are so troublesome for studies which claim to establish the absence of effects or hazard that such studies will hereafter be explicitly excluded from consideration.

The purpose of the data-processing operations is simply to facilitate the next step, the biostatistical analysis that will go on in the black box. For example, the computerized operations which produced the observations in Figs. 5 and 6 in minutes at one time took us months (when the tabulation had to be done manually by clerks). Operations in statistical tabulation which select or extract information are generally regarded as admissible, although there may sometimes be some questions as to whether all of the relevant information was obtained. In general, the specific linguistic transformations used here are those of the mathematical languages and, in particular, of arithmetic.

In going from Figs. 5 and 6 to the estimate of the doubling dose, a sophisticated analysis was used that had been developed specifically for the purpose. In this sense the technology was new. However, the procedure really consisted of putting together components which individually were old and familiar ones. This will nearly always be the case, so the strategy for determining admissibility consists of identifying the components in the analysis and assessing the admissibility of each component. Only rarely would it be necessary to deal with procedures which are entirely new.

Indeed, if one wants to pursue this "divide and conquer" strategy for understanding the machinery of a scientific argument, the data-analysis procedures can be broken down further and further. Eventually, there is nothing but very simple operations that would be called "arithmetical" or "clerical." These are the kind of elementary operations that can be performed by very young children in school. They can also be carried out on an electronic computer since they are further reducible to moving "bits" of information around in a magnetic memory.

The performance of these pure linguistic operations requires no knowledge, judgment, or common sense. It can be done very fast and very accurately on computers. This is where computers are useful. Translating these simple operations into programming language is where computer programmers are useful. But whether the output of the computer is going to be scientifically meaningful or merely useless printout will depend on the higher-level instructions. Despite some pretentious claims, neither computers nor programmers nor computer experts have much understanding of the scientific strategies that are involved. All of the computer operations, if done correctly, simply carry out someone's instructions. If that someone doesn't know his or her business, the computing is a waste of time and paper.

JUDGING THEORETICAL INPUT

Some readers may have been waiting patiently for some place in this discussion of the machinery of a scientific argument where genius, creativity, brilliance, or other higher intellectual activities would come into the picture. Up to this point, the operations described have involved such pedestrian virtues as care, forethought, hard work, or persistence. The operations themselves were rather routine or reducible to routine operations. While management of these operations were likely to involve some practical experience in scientific research, they have not ordinarily involved any great breadth or depth of scientific knowledge. So some readers may be asking: Is this all there is to a scientific argument? Is there no glamour and no glory?

The truth is that there is little glamour or glory in good sound scientific research. The spectacular breakthroughs and other media promotions are little more than the kind of fakery that is nowadays considered "good public relations." Edison's estimate of 99 percent perspiration and 1 percent inspiration is not far from the mark when we are talking about genuine research or valid scientific arguments.

It should also be remembered that we are discussing scientific arguments in the context of justification. Whatever brilliancies may have been involved in the design of the study, the development of the theory, or the evolution of a technology for rigorously comparing theory and fact will show up in the results rather than in the structure of the argument. It is the soundness of this structure that is being evaluated, and no points are given for the cleverness or ingenuity of the original conceptions.

If there is anyplace in the black box where there are any vestigial signs of any earlier genius in the study, it is probably in the component labeled "theory." The theory will get the same rigorous testing whether it represents a deep insight or flash of genius or prevailing "conventional wisdom." The difference between sensible and foolish theories only becomes evident at the very end when one is likely to be accepted and the other rejected. Thus, the proof of the pudding only comes *after* the long and arduous process of building a sound scientific argument (plus any brilliance or luck that may have indirectly helped in this process), which eventually leads to new and fruitful findings.

In this sense, then, the creativity that may have been incorporated in the theoretical input is reflected in the "output statements" and not in the black box itself. For instance, while the formal statement of the theory under test at the NRC hearing involves a new and important concept of "co-occurrence," there is little in the algebraic statements of the theory to indicate that this was the case. A foolish speculation could be expressed in mathematical formulas that would look very much the same. Someone familiar with the scientific context might be able to spot the point, but this would not be apparent to a concerned citizen, to a mathematician, or to a scientist in the radiation protection community.

This detachment of a theory, particularly a theory expressed in algebraic language, from the realities of a scientific context might be called "autonomy." This autonomy is both good and bad. It is good because it allows the expression and testing of ideas which might otherwise be rejected out of hand on the basis of the prevailing "conventional wisdom." On the other hand, it makes it difficult to have quality control on the theory going into the black box that is as tight as that on the data. However, the following schematic suggests that some quality control can be achieved:

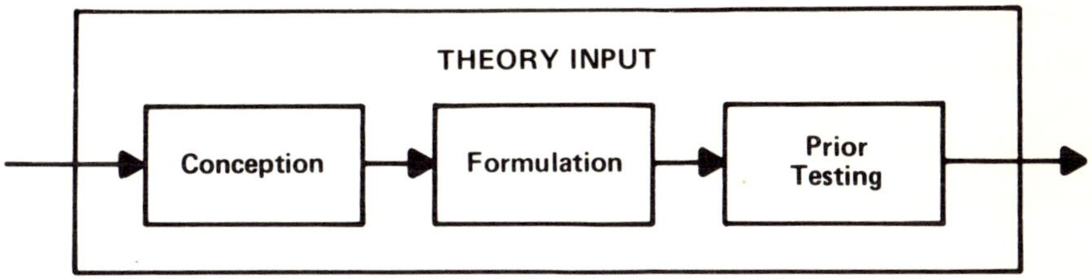

Theory input.

Note that while scientific knowledge or brilliant intuitions or operational experience can come into the black box through this route, they enter with a somewhat different status than the data input. The theory that enters has only a *tentative* admission. It can be completely rejected at a later stage.

The data that is admitted has a stronger position. It is not accepted absolutely. In the subsequent operations in the black box, an apparently aberrant data point may be excluded. For example, this type of post hoc exclusion gives rise to the dotted curve in Fig. 4. The estimate of the doubling dose is somewhat reduced by eliminating some seemingly aberrant data. But this rejection involves only a few data points rather than a general rejection of all the data.

In comparing the status of the data input with the theory input to the black box, the linguistic structure of the statements helps to make this distinction. Theory consists of statements in the form "if-then." In other words, the statement is conditional and of the form "If such-and-such is true, then so-and-so will occur." In contrast, the data statements are simply reports of events in the world about us. They are asserted unconditionally.

In this sense, then, it isn't really necessary to judge the theoretic input. This judgment will be made by objective, scientific techniques at a later point.

THE REALITIES OF THEORY

There is some tendency for those in the academic world to greatly admire themselves and their own works and to depreciate everyone else. For the most part, the academic world is engaged in talking rather than doing. It does teaching-type "research" which produces dissertations rather than scientific findings. There is a stress on impressing students with the proper respect for the importance of their disciplines (and their teachers). This leads most students to get a badly distorted picture of science that may hamper them for years or for life. Theorists such as Einstein are glorified, while bench scientists such as Edison are dismissed as rude empirics. It is part of a general academic pattern which extols clever manipulations of language above all else.

What is striking is this: Very few of the people who adulate theory have ever made any substantial contribution to the development of a new scientific theory. The most they have ever done is to pretty up some existing theory. So academic theory-appreciation courses show little understanding of the *realities* of theory.

The whole idea that theory is something esoteric, elevated, and apart from grubby facts is directly contrary to the scientific reality. As might be expected from the Galilean rule, an intimate and continuous contact with data is essential to the development of effective theory.

To give the reader a better idea of the realities of theory and to make the role of theory in a scientific argument somewhat clearer, let me consider briefly the historical development of the theory that was involved in the NRC presentation.

Among the underlying theoretic ideas that have been circulating in one form or another for the past decade or two are the notions that genetic instructions for cells are carried in DNA, that these instructions are used to produce the enzymes and other chemicals needed in cellular or organ systems, and that these codes can be disrupted by structural damage to the DNA in the genetic material. If these notions are combined with the well-known fact that low levels of ionizing radiation are capable of producing permanent damage to the structure of the DNA in genetic material, then a theory that radiation exposures could produce multiple failures in the body systems of the person exposed (or a child of the person) years after the exposure seems quite plausible.

By the late 1960s, the possibility of a co-occurrence hypothesis had already occurred to me. It seemed likely that radiation could produce more than one adverse health effect in the exposed person. However, there were a number of competing speculative notions that also seemed reasonable. The existing statistical methods were poorly suited for distinguishing between specific rival theory because they are general-purpose, nonspecific methods. Since I didn't see how to put this co-occurrence theory to a rigorous empirical test, it remained an "interesting speculation" until well into the 1970s.

This experience with theory is not, I believe, at all exceptional. It is a "fact of life" with theory that good ideas are rare. However, it is even rarer to get good ideas which can be formulated with sufficient precision to put to a rigorous test. Finally, only in very exceptional circumstances is good data at hand that can give a rigorous test of the theory.

Formulating a theory in precise mathematical language is sometimes difficult in itself. Much has been said on this subject by those who emphasize the role of mathematics. But in building a successful theory, what is even more difficult is to avoid going off on theoretical paths which lead only to dead ends or computationally hopeless problems. For instance, the simple mathematical model of Fig. 9A and 9B is not the theory I started out with.

My original model directly involved the number of breakpoints in the DNA of human genetic material and the effects of radiation in producing further breakpoints. This was a much deeper model than the one in Fig. 9A and 9B, and it also involved many more unknown parameters. For instance, it involved the distribution of the breakpoint count in the human population, the tendency for this count to increase with time (or age) due to cosmic rays and other environmental mutagens, and so on. The simple model was developed as a first approximation to the complex model. Its purpose was to help us get some rough estimates for the unknown parameters of the deeper model.

When theory is evolved in close proximity to data, there are always surprises--some good and some bad. Most surprises are unpleasant, but what happened when the simple model was used came as a pleasant surprise. Despite the inelasticity of this three-parameter model, it fitted a small piece of the childhood leukemia data (one age group and exposure category) very nicely. It was obvious to me that the model was much too rigid to have much chance of surviving further tests, but I thought it would be instructive to see how it failed. So Mr. Natarajan and I kept trying it out on more and more data. The model kept working. We seemed to be onto something.

There was to be another long delay before we could test the model on the adult data. Our original fitting procedures were only partially automated and involved quite a bit of cut-and-try work. The amount of time needed was excessive. Fortunately, Mr. Steven Falen came along and put together the necessary software to do the job in push-button style.

One hidden advantage of this delay was that the "prior testing" component of the schematic which provides the quality control for the theoretic input had been carried out fairly extensively on the childhood data before we looked at the adult data. But this itself created a potential problem.

When you work with a model for a while, particularly when it is your brainchild, it is only natural to develop an attachment to it. If the Galilean rule then leads to a "scientific tragedy" and the theory is killed by the facts, then this emotional attachment to a theory can be very painful. I was afraid that this might happen to us since we had worked with the model for so long. I try to avoid this kind of emotional attachment to any mathematical model, but it isn't something easy to control.

I didn't really expect the model to work on the adult data, so I was pleasantly surprised when it did. This carryover of a model to entirely new data is not often successful. There are major differences here between adult and childhood leukemias. To complicate matters, dosimetry comes into the picture in the adult data, and this is always a can of worms.

When the simple model continued to describe the Tri-State Survey data, this forced me to reexamine the notion that the simple model was merely a useful approximation to a deeper model. It now seemed that some essential features of the deep structure to the action situation were being characterized by this model. This reconsideration is an example of feedback from the testing to the conceptual stage, but is an unusual kind of feedback.

THE THEORY GAME

Perhaps the previous section may serve to make the point that the realities of theory are not particularly glamorous. The vicissitudes of developing a successful theory are not so very different from those of developing good data. For best results, the two should constantly interact with each other.

The scientific realities of theory should not be confused with what might be called the "Theory Game." This game is, like most academic games, essentially a language game. I have been describing what is involved in *doing* theoretical work. However, the main activities in academia is *talking* theory. Indeed, there is likely to be more talk about theory than about any other aspect of a scientific study (with the possible exception of some favorite technical points).

The Theory Game, like many others, is essentially a form of one-upmanship. It inflates one's ego and self-importance at someone else's expense. Often, the Theory Game is played in lieu of genuine criticism. Whether it is at the peer review of a grant proposal, the review of a scientific paper, or in other decision-making, playing the Theory Game allows one to speak "in the name of science" without knowing or doing any real science.

Perhaps the schematics and discussion in this chapter may help the reader to debunk this Theory Game. An extended discussion of a theory which is going into a black box is, as we've seen, unnecessary. The theory goes in with a question mark automatically attached. It is a conditional "if-then" statement. It will be the data, and *not* the opinions of the players of the Theory Game, that provides the acid test of the theory.

The whole point of insisting on the objectivity of the operations in the overall scientific argument is to avoid subjectivity and personal biases. In deference to the sensibilities of scientists, it is customary to say that these precautions are needed because interviewers or technicians or nurses might be biased. The truth of the matter is that these procedures are mainly self-discipline for scientists themselves. They are the ones who are likely to be the most opinionated, most biased, and most subjective persons of all, because they are the ones who become identified with particular theories.

chapter 9

SOME PRACTICAL STRATEGIES IN BIOSTATISTICS

Elements of strategy and techniques in the statistical analysis of biomedical data.

INSIDE THE BLACK BOX

Over the years I have come to the view that most of the troubles of our technological societies are a result of confidently asserted and completely fatuous "expert opinion." I had an early warning of this when I noticed that the conventional wisdom in the medical world was more often wrong than right. Later I realized the corollary of this principle: The main reason we have done so poorly in finding remedies for the troubles of our technological societies is that no sensible recommendation can escape the criticism and last-ditch opposition that is predicated on authoritative but erroneous expert opinion. It is not just the ignorance of experts (which they will never admit) that creates the difficulties; it is the arrogance that goes with this ignorance.

So perhaps the most important talent that you need to save your life nowadays is the ability to distinguish between the handful of real experts in any area and the swarms of phony experts in these areas. If you can't tell the difference, at least you can learn to doubt the opinions of experts (particularly when no factual backup for the opinion is forthcoming). The more authoritatively the opinion is asserted, the less credence you should put in it.

It may be overly optimistic on my part, but I believe that if you have some understanding of what goes on inside the black box that I have used to represent a scientific argument, you will be better able to distinguish between experts who are using a genuine argument and those who are simply snowing you with a lot of technical-sounding double-talk. So in this chapter we will pry the lid off the components in the black box and see what goes on inside.

This is not as difficult as it might seem to be. We are always going to be dealing with linguistic operations. These operations of scientific languages will always have counterparts in the operations of everyday English.

OPERATIONS ON DATA

In Chapter 8 we have taken a look at what goes into the black box by way of data and theory, and at how the components of the black box are organized to put the Galilean rule to work. Now we will take a closer look at the operations involving data and theory that go on inside the black box. Once again, the NRC presentation on radiation hazards can provide a concrete realization of the abstract conceptualizations of what goes on in the black box. We will take a closer look at the operations on data in general, and the Tri-State Survey data in particular, that are carried out inside the black box that represents a scientific argument. In this way we can get a better idea of how the information is compressed, how half a million statements in the input can be compressed to half a dozen or fewer scientific findings in the output.

To get us started, we can again consider a schematic of the process:

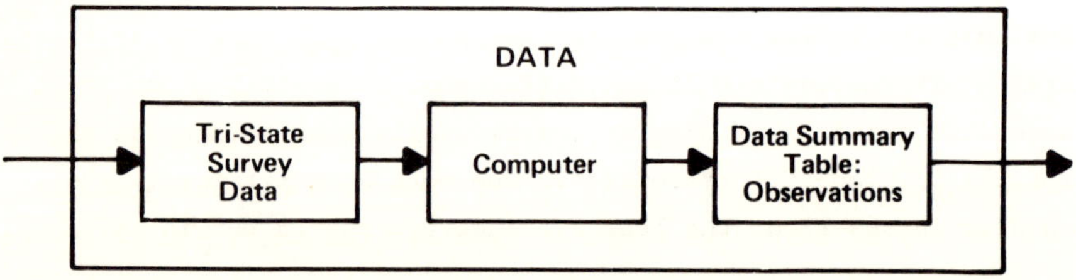

Data for black box.

The information coming from the Tri-State Survey can, for concreteness, be regarded as an enormous number of magnetic symbols on a computer tape. The summary tables are exemplified by Figs. 5 and 6 for the men (or Figs. 11, 12, and 13 for the children). Here information abstracted from the tape is presented in the form of an array of numerical counts which were previously referred to as "observations."

The black box inbetween the data tapes and the tabulations is labeled here "computer." This provides a concrete realization of how the actual operation in the study (and most studies nowadays) are actually carried out. Putting the "computer" box into the diagram is not a way to emphasize the importance of the computer in the overall process of a scientific argument, but rather a way to *deemphasize* it. The point is that any operations that can be carried out on a computer must be purely linguistic operations, which are completely reducible to a mathematical operation of turning a switch on or off. In other words, these operations can be carried out in a completely mindless and mechanical way. A more impressive way of saying the same thing is to call them "algorithms."

Any concerned citizen--or for that matter any scientist--who is awed or impressed by any operation that can be represented by a box labeled "computer" is simply gullible. Anyone who points to such operations as evidence of the "infallibility" of an argument, or tries to glorify computer operations (e.g., by talking about "machine intelligence") is selling you a bill of goods.

A computer is a tool. That is all that it is. Whether the printouts are informative or nothing but waste paper doesn't depend on the machine. It depends on the persons who are using the tool. If they don't know what they are doing (and my experience has been that much of the time they don't), then what comes out are false and misleading numbers. If technologists insist that the numbers are realistic and if they succeed in deploying potentially dangerous technology on this basis, then, in this sense, a computer can kill you. More precisely, the computer is a neutral tool, and it is the mindless use of computers by human beings that can kill you.

The reason that the "computer" box has a useful role in the diagram is that it serves to emphasize that the tactical or technical operations are reducible to a mechanical routine. So while statisticians or epidemiologists or other experts love to dwell on these "safe" technicalities, they are not what is really important in a scientific argument. Mastery of these technologies may be important to someone who is doing a study. It is not important to the concerned citizens or anyone who is trying to understand the scientific argument. It is the strategic ideas involved in the operations in the previous diagram that matter, and these ideas can be understood in common sense terms and plain English.

THE REASON WHY

To consider the data operations from a strategic point of view rather than a purely tactical one, we can start by looking at the tables that are the end product. We can see that Fig. 5 is limited to certain kinds of information. Thus, the information here involves dosage of radiation, age of the man, and health status with respect to nonlymphatic leukemia and heart disease. This is a small fraction of the total information on the Tri-State data tapes. Hence, much of the compression has been achieved by extracting subsets of information from the total set of information. This strategy for compression is habitually used in ordinary discourse where, for instance, an eyewitness report of an event cannot include every impression. The reporter selects what is relevant or interesting or useful for the discussion.

In understanding the data operations in the black box, there are a large number of questions that might be asked. We will consider a few of the more important ones here. For any phase of a scientific argument, it is important that the person carrying out the operations knows what he or she is doing. This is particularly important in selection operations. Anyone who knows what is going on can give sensible answers in plain English. However, people who run data through canned computer programs don't really know what goes on when they push the button. So often they resort to technical-sounding evasions for answers.

Why, then, were the dosage and age and other items of information in Fig. 5 selected out of the total information in the original data tape? The general answer is that there was a specific theory under test. This involved these particular items and not the others. While this answer is generally true here and in other scientific arguments, there are some specific selection decisions which are not covered by this umbrella answer.

For example, this is data where the focus is on nonlymphatic leukemia in men. Why nonlymphatic leukemia? Why men? This choice was made to give the clearest test of the co-occurrence hypothesis (and to get the clearest dosage response curve and doubling dose estimate). But why this particular choice? The answer involves past experience both with biostatistical studies and with the scientific context of this study. This is an area that neither concerned citizens nor scientists in the radiation protection community can be expected to know very much about. Indeed, the number of experts knowledgeable on these matters is very limited. There are probably not more than half a dozen research centers where this kind of expertise is available.

First, why men? Why not women? This isn't a sexist decision. In leukemia and in a number of other diseases the relationships to radiation and certain other factors show up much more clearly in men. There are other diseases where a study in women is likely to be more conclusive.

Why not study both? We planned to start with men where things would be clearest and then go on to study women. This plan ended when our funding ended. Why not study both men and women together? A full answer involved technical points that won't be elaborated here. The basic idea is this: If there is a subseries with strong effects and a subseries with weak effects and they are put together, the effect is to dilute the strong effects and sometimes to lose them. This is why it is a key strategy of epidemiological research to look for subseries where the relationships will be clearest and strongest. In those epidemiological studies where we are working at a research frontier, we are often working close to the limits of the resolving power of our statistical instruments. So unless we make wise analytic choices, we won't see anything at all.

In addition to the strategic reasons, there is a tactical reason for this choice. It involves presentation and publication. The prevailing view among physicians is that radiation affects men but not women and that it does not produce lymphatic leukemias. This conventional wisdom is probably wrong. However, it eases the task of presentation if one can use material which will be more acceptable to the audience, because it doesn't conflict with all their preconceptions at once.

Since these radiation studies encounter fanatical resistance, it is tactically unwise to create unnecessary targets for objections. The conventional wisdom derives from data involving much higher dosages. It might not be true in the 1-rad range. The current opinion of most radiologists is that the low doses were essentially harmless, so the study was bound to create a furor when it challenged this article of faith. This is the central issue, and it is controversial enough in its own right. Hence, it was good tactics to avoid unnecessary side issues and to deal with nonlymphatic leukemia in men.

STRATEGY AND TACTICS

When tactical and strategic purposes agree, this reinforces the choice. When they disagree, the choice may be a hard one. Let us start with a case where the tactical choice and the strategic choice were in agreement. The decision to use dosage categories is a good example.

A natural question for readers who are quantitative-minded would be: Why did we group dosages into dosage categories instead of using the numerical estimate of the dosage? Doesn't this waste information?

At first sight this may look like a situation where the tactics conflict with a sound strategy (don't waste information). However, the strategy and research experience from one area can sometimes be misleading in a different area. When one deals with genuine measurement data (such as may be found in laboratory experiments in the physical sciences), it would be pointless to lose information by using dosage categories instead of the actual numerical dosages. But the numbers in dosimetry are not really measurements; they are calculated from other numbers.

For instance, if a certain radiological procedure is listed as 2 rads, this number may have originally been a measurement in a laboratory setting. But when this number is being applied to x-ray delivered in the field, it is not a measurement. The actual dose can vary widely. Thus, the mammography dosage of 2 rads which was measured in the lab was also checked out in the field. Here the actual dosages ranged from a (claimed) 0.2 rads to 6 or 8 rads. There were even a few double digit figures.

In biomedical research, the numerical values frequently have properties that are very different from those of genuine measurements. Hence, the numbers for radiation doses have a kind of false precision. It is more realistic to use statistical methods which don't make any assumption that the numbers have the nice statistical properties of true measurements. Dealing with dosage ranges rather than numerical dosages is one simple way to avoid strong assumptions. This gives what is called a "more robust" analysis. It is well worth trading a little loss of information for a more reliable conclusion. This is a good strategy and good tactics in a biomedical study.

As you can see, there are various questions that can be raised about the selection of a subset of a data set. There are sensible, nontechnical answers that can be given to these questions. This is largely because it is the strategic purpose rather than the statistical details that really matter. However, in some cases the selection decisions are forced by tactical considerations and the peculiarities of a given study situation. Even there, however, the explanation can be in common sense terms rather than in technical jargon.

One example of a choice that was more or less forced by the tactical situation is the choice of heart disease as the co-occurrent disease. In part this choice was based on earlier studies with Dr. Enrico Viadana where we found that heart disease was predictive of leukemia. We had wanted to go further and look into radiation effects at the time, but unfortunately standard statistical methods couldn't do this job. The limiting factor is the empty or near-empty cells of the contingency tables that occur when there is too much cross-tabulation.

In addition to heart disease, we found several other diseases predictive of leukemia. Indeed, some were more predictive than heart disease. So why did we use heart disease?

The technical feature that tended to force this decision was sample size. Heart disease was much more common than the other diseases in this age and sex subseries. While analysis using the mathematical model is less vulnerable than standard techniques to empty cells in the contingency table, this was still a limiting factor. At that point we didn't know whether the relationship between heart disease and leukemia would also be connected with radiation exposure. However, the larger series would make it easier to look into the possibility and give a better chance of seeing an effect if it were there.

Another choice that was forced on us was the omission of the adult cases where no radiation was reported. This wasn't a decision that was made for this particular study. We had been forced to do this in our previous studies. The problem here goes back to the data-collection stage. With children, there was generally a knowledgeable respondent such as the mother. With adults, particularly when the interview could not be carried out before the death of the leukemia victim, the closest relative was sometimes not the spouse but a more remote relative. In these circumstances, there were two different reasons why there might be no report of x-rays. On the one hand it could be because there was no radiation. On the other hand it might be because the respondent didn't know much about the medical history of the leukemia case. Previous studies showed that the latter was frequently the explanation. Hence, this was a "mixed" exposure series and we could not "unmix" it. Over the past decade, we had found this subseries was not informative and that including it only confused matters. So we left it out.

A third choice involves age and the choice of particular age groups. In this instance, the inclusion of the age variable is virtually forced by tradition. In cancer studies, the increasing incidence with age is such a strong relationship that if it is not controlled in some way, any effects for another variable such as radiation might be attributed to some kind of age bias. The particular age groups used here are in part determined by the age stratification in the sample, and in part by the conventional use of round numbers for categorization of continuous variables.

Incidently, there is a good reason for this traditional use of "round numbers." Otherwise, the scale might be retrospectively divided up into categories in such a way as to favor an investigator's hypothesis. The use of round numbers somewhat restricts this game. This is a good example of a seemingly arbitrary linguistic convention that really has a lot of sense behind it.

OPERATIONS ON THEORY

Let us turn now to the black box labeled "theory" in the black box representing the scientific argument. We find operations somewhat parallel to the "data" box when we take off the lid of the "theory" box and look inside:

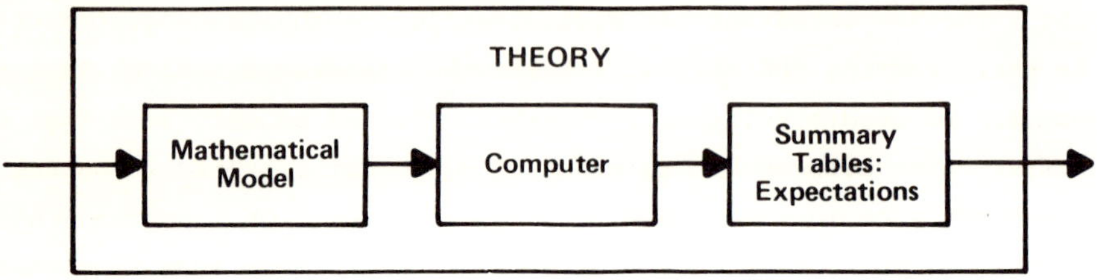

Theory for black box.

This time the operations are on numbers which are generated by the mathematical model. The summary tables that result are predictions or expectations. These are the numbers that will be compared with the data counts when the goodness-of-fit is evaluated. For theory as for the data, a computer box is the middle box. Hence, here too, the black-box operations are reducible to elementary linguistic operations which can be carried out in a mindless way.

As before, the strictly operational or technical matters are not the ones which are important from a scientific standpoint. The technologists who are involved with them, whether mathematicians or computer scientists, like to make an elaborate production out of these intellectually trivial operations.

The questions that are important to an understanding of a scientific argument do not concern the mechanics of these low-level linguistic transformations, but rather involve the question: "Why are these operations being done?" As before, the answer to this question at the strategic level is the key to understanding the process. Once again it can be given in common sense terms and in plain English.

Here, as in any genuine scientific study, the role of the mathematical model is to describe events going on in the real world. In particular, its role here is to describe the "observations" that are the output of the data operation. Whether or not the model is successful in this role will be determined later when the operations in the "fit" box are carried through. The operations in the "theory" box are designed to set things up for this eventual acid test. The "expectations" in Figs. 5 and 6 are supposed to predict the corresponding "observations," and the mathematical model is supposed to generate these predictions. As can be seen from the figures, even without the machinery for determining fit, the model does quite well.

What about the mathematical model itself? What does it mean in terms of the scientific context of the study? The algebraic formulation of the model for the children in Fig. 9A and 9B gives a precise specification of the model in a form which can be readily translated into a computer program. However, it may not mean much to most readers. Contrary to what some readers might suppose, this *isn't* because of their mathematical deficiencies.

I had a long and fruitless correspondence with a British physicist who simply could not understand this model. The mathematics here are elementary, even for a physicist. However, the process of linking abstract symbols to the real world is something which surprisingly few scientists have had much experience with. You can use the mathematics in a textbook or handbook without having to make this kind of linkage. It can be done in a mechanical way. However, if you are doing new research rather than teaching a fossilized form of science, then the ability to make this kind of connection between the model and reality is essential.

THE SAMPLING MODEL

The model for the childhood leukemia is similar to, but somewhat simpler than, the model for the men. In my NRC presentation and here it provides a better illustration of the role of the model. We can postpone the complications produced by the radiation dosage for the time being. The first point to note is that there are two more-or-less separate and independent models that are being used here, and this will generally be the case (although one of the models may come only in the degenerate form of an "error" term). The schematic of the mathematical model would be:

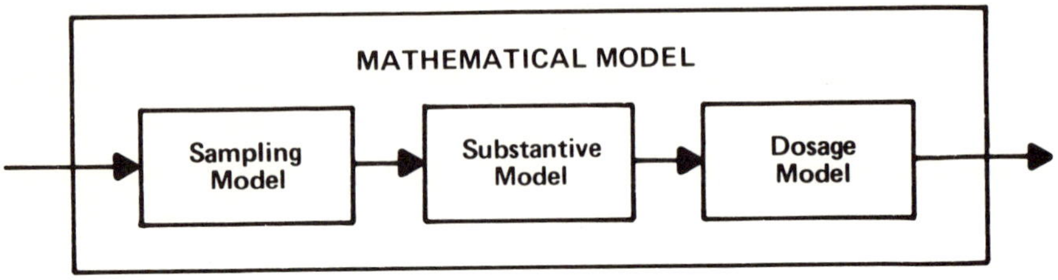

Mathematical model.

The sampling model simply describes how the Tri-State Survey data were generated. This is called the "model" in Fig. 10. The substantive model is called the "adjusted" or "extended" model in Figs. 10 and 11. This idea is very simple and the representation of the idea is also simple. The probabilities and expectations are all calculated by repeated multiplications.

It is worthwhile to go through one specific example simply to show that it is the unintelligibility of technical languages rather than complexity of scientific ideas that creates the apparent difficulty. So let us start with the formula in Fig. 9A:

$$P_0(X_1,X_2,X_3) = [D_0(X_1)]^{X_2}[1 - D_0(X_1)]^{1-X_2}[L_0(X_1) \times K]^{X_3} [1 - L_0(X_1) \times K]^{1-X_3}$$

This may seem formidable, but it is simply a matter of an ingenious mathematical notation for very compact representation--a purely linguistic device for another kind of compression. The basic idea here is to write one formula instead of a lot of separate formulas. This is accomplished by the use of an ingenious but unimportant linguistic device sometimes called "dummy exponents."

First of all, consider the variables symbolized by X's. One of the variables (radiation exposure status) does not appear in this formula because this is the formula for "Case 1: No Radiation Exposure Reported." The other variables are age (X_1), occurrence of indicator disease (X_2), and occurrence of leukemia (X_3). The reason for using the subscripts (such as the subscript 1 in (X_1) is notational economy. This saves having to use three different letters.

Letting X_2 (and X_3) be equal to zero if the disease is not reported and to 1 if the disease is reported is a device for killing two birds with one stone. On the one hand it provides a useful way to assign numbers to categories for computerized operations. On the other hand, the trick with the dummy exponents compresses the presentation of the formulas.

The trick itself is simple once you see it. It relies on the fact that any quantity raised to the exponent zero is equal to 1. Any quantity raised to the exponent 1 is simply that quantity. If Y is any nonzero quantity, the familiar formulas are:

$$Y^0 = 1 \qquad Y^1 = Y$$

What this means is that the terms in the formula involving status with respect to indicator diseases can be switched in or out by using the exponent. In other words, we want $D_0(X_1)$ in the formula when there is a report of indicator disease so we write:

$$[D_0(X_1)]^{X_2}$$

Therefore when $X_2 = 1$, the term $D_0(X_1)$ comes in. When $X_2 = 0$, this term goes out and the term

$$[1 - D_0(X_1)]^{1-X_2}$$

comes in. In this way, a single formula covers all of the no-radiation cells of the contingency table. Thus, if we want to consider a specific cell, say, the cell where both indicator diseases and leukemia are reported, the formula simplifies to what we want. Thus, setting $X_2 = 1$ and $X_3 = 1$:

$$P_0(X_1,X_2,X_3) = [D_0(X_1)]^1[1 - D_0(X_1)]^0[L_0(X_1) \times K]^1$$
$$[1 - L_0(X_1)] \times K)^0$$

$$P_0(X_1,1,1) = D_0(X_1) \times L_0(X_1) \times K$$

If, for instance, we further specify the youngest age group ($X_1 = 1$), then we see that P_0 is the product of three specific numbers:

$$P_0(1,1,1) = D_0(1) \times L_0(1) \times K$$

What this equation says, in ordinary language, is that the probability of a positive report for both disease conditions is the probability of leukemia $[L_0(1)K]$ times the probability of indicator diseases $[D_0(1)]$ in the age group. The leukemia probability is the probability in a single year $[L_0(1)]$ times the number of years of the study (K). For the probability of not getting leukemia there is a slight approximation in using $[1 - L_0(1)K]$.

If the disease risks are "independent" in the statistical sense, then this multiplication of probabilities is valid. The test is to see what occurs in the data. For the nonradiated series the independence assumption predicts the data.

The expectations involve some further multiplications. The number in the population in this age category $N(X_1)$ is multiplied by the proportion reporting no radiation exposure $H_0(X_1)$ to get the numbers of children in the population with no radiation exposure. For the random sample ($X_2 = 0$), this is multiplied by the sampling rate (S), and otherwise not.

Sampling models simply translate algebraic language into a description of the sampling process used in a study. After that has been done, it is just a matter of arithmetic to calculate the expectations.

The random sample procedure in the Tri-State Survey is what makes this simple and straightforward mathematical model possible here. Very few epidemiological studies have this resource because there is no random sample. For most, there is no population base either. Once you are dealing with a random sample and a complete enumeration (for the leukemic households), you can write down this multiplicative mathematical model immediately. This doesn't mean, of course, that that particular model is going to predict the actual observations (although it might be expected to). This still has to be checked out using the random sample data and the children who were in the leukemic series but had no report of radiation exposure. This was the first thing that was done and the prediction was excellent. However, this was only the first step.

THE SUBSTANTIVE MODEL

Once the sampling model checks out, we can go on to the substantive model. The substantive model here characterizes the relationship between exposure to radiation and risk of subsequent leukemia and other diseases. Again, both the idea and the algebraic statement of the co-occurrence hypothesis are simple and straightforward.

The central concept is that when a large number of individuals are exposed to low-level ionizing radiation, only a small fraction of those exposed are affected in a detectable way. In other words, certain effects of the radiation could be detected in the Tri-State data while other effects could not. For instance, effects which produced death in the first year of age or earlier cannot be detected, since such deaths could not come into this study. The main scientific point is that these affected individuals come from a susceptible subgroup of the population. This crucial point will be considered further in the next section. Mathematically, however, all that is involved is the introduction of a parameter, A, to denote the proportion affected by the radiation exposure.

Once the little trick with the dummy exponents is obvious, then the probability in the youngest age series with both indicator diseases and leukemia is reported as a simple multiplicative form becoming:

$$P_1(1,1,1) = D_1(X_1) \times L_1(X_1) \times K$$

Letting R_D be the relative risk in this affected group for the indicator diseases and R_L be the corresponding risk for leukemia, then by definition:

$$D_1(X_1) = R_D \times D_0(X_1)$$

$$L_1(X_1) = R_L \times L_0(X_1)$$

The situation in the affected series is therefore characterized by the two relative risks. We might anticipate that R_D and R_L are going to be larger than 1, but we leave it up to the facts to freely determine what the relative risks will be. For those individuals that are not affected, there is no change in risks from the formulas in the original model where no radiation exposure was reported.

The expected number is therefore calculated by addition. The expectation for the proportion (1 - A) of the group that is not affected must be added to the expectation in the proportion of the group (A) that is affected in order to get $E_1(X_1,X_2,X_3)$. Note that saying what we expect in algebraic language is not really so different from saying this in ordinary language.

As you can see, the algebra which might initially seem like the hard part of understanding a mathematical model is not really a serious obstacle, even to a nonmathematician. This kind of mathematical manipulation is easily checked, and you can rest assured that if any slips are found here, they will be trumpeted all over. So even when we are dealing with a mathematical model, the strategic ideas are what matter.

For instance, the crux of the co-occurrence model is that only a relatively small proportion of the persons exposed to x-ray are affected by it in ways which are detectable with this data and technology. The "proportion affected" is not a mysterious or abstruse idea, and its symbolic representation should cause no problems (even though it seems to baffle the entire radiation protection community).

SUSCEPTIBLE SUBGROUPS

This is a rather important question, not just here, but in the investigation of the hazards of chemical carcinogens and other environmental factors that produce chronic diseases. In general, across-the-board effects are rare. It is some small fraction of vulnerable persons in the population who are affected. If we can't get a handle on this susceptible subgroup we are likely to miss the effect or misunderstand what is happening.

We are dealing with especially vulnerable persons when we look at the side effects of anesthetics such as Halothane or of antibiotics or other medications. A substantial part of all adverse side effects, of all iatrogenic disease, comes from these susceptible subgroups. For any one chemical or any one carcinogen, however, the proportion affected may be very small. *The key point is that this isn't a random subgroup.*

A weakness of most current statistical analyses is the tacit assumption that the side effects occur in a random subgroup of the general population. This is, for instance, one of the faulty assumptions that leads to the failure of methods for setting "safe" dosage levels that use linear extrapolation. The affected persons have something special about them, probably preexisting defects in their genetic material.

The existence of these susceptible subgroups is what makes it so very difficult to provide effective protection against hazards in the environment. If the protective levels are set to "protect" the general population, they don't protect the susceptible subgroups in the general population.

Learning more about these subgroups and how to detect them is a scientific challenge in public health. This is the motivation for putting this notion into a mathematical model of radiation effects. It takes time and effort to try to work out a mathematical model involving a susceptible subgroup and to fit the parameters, the proportion affected, and the relative risks of the diseases in this affected group. It is much easier to run a canned statistical program on the data and to end up with nothing at all, as was done in the "critiques" of this work).

However, if we really want to protect the public we cannot do the work in this slipshod way. If we want to, say, protect women against potential carcinogens in hair dyes, it does little good to grind through routine analyses of the data that will miss the hazards completely. Thus, to find out what is really going on with hair dyes and breast cancer, it is essential to see what happens in the high-risk groups. When scientists look there they find the effects.

Unfortunately, we are in a situation where the federal agencies reward peacefully negative results. So nowadays the "smart" thing to do is to carry out an incompetent statistical analysis of data. This can easily be accomplished by using the canned programs that are widely available instead of a statistical analysis that is appropriate to the data. Then the results will be comforting to an industry and may even be comforting to the victim (until, of course, the disease actually occurs).

While the underlying conception of subgroups is incorporated into the co-occurrence model, the underlying concepts are not very visible in the mathematics. Nor is there anything in the algebra that explains why heart disease has come into a research study where the focus was on leukemia. The mathematics specifies the relationship. It doesn't explain why we want to study this relationship.

The algebraic statement is like an abstract general idea that can only become useful when it is connected with a concrete application. The abstract notion that the same cause can have a multiplicity of effects is useful only when it is connected to a specific cause (e.g., exposure to x-rays) and specific effects (e.g., heart disease and leukemia). Then it can provide a more powerful instrument for studying radiation effects in diseases where radiation may be a relatively minor factor. It can enhance the resolving power of our statistical instruments and enable us to pick up effects at dosages where the proportion affected will be very small. That is what the mathematics is all about. However, none of this is obvious from the mathematics itself. It can only be noted when the mathematics is considered in the scientific context of the study.

chapter 10

SOME MATHEMATICAL STRATEGIES IN BIOSTATISTICS

Elements of strategy and techniques in mathematical models for public health problems.

SECRETS OF A MATHEMATICAL MODEL

How we can better understand the role of theory and what goes on when a theory is expressed in mathematical language (and is called a "mathematical model")? We can start from Bertrand Russell's sage observation that a mathematician is someone who doesn't know what he is talking about.

Russell was making a little joke about this vital point: When a mathematician deals with abstract symbols, it does not matter what the connection might be between the symbols and the real world (or even whether there is any connection at all). This detachment from reality is sometimes a considerable advantage. It enables the mathematician to carry out complex manipulations of his symbols without having to worry at every step (or at any step) just what it all means in terms of reality.

On the other hand, if you don't know what you are talking about, you can say some crazy things. In the physical sciences, this freedom from the chains of reality, this license to explore the world of "if-then" or "what if?", has led to some surprising insights. However, it can be a very dangerous path to follow in health issues. If you lose contact with reality, it is easy to lose your footing on these slippery paths of linguistic abstractions. You can then end up saying something that is flatly false, although you are laboring under the delusion that it is absolutely true. This has, in fact, happened over and over again with mathematical models which were supposed to predict radiation hazards.

I have had an opportunity to observe the contributions of mathematicians in various areas of the sciences, particularly in such areas as statistics and linguistics. I have reluctantly come to the conclusion that the use of mathematical languages in a discipline is far from an unmixed blessing. In one discipline after another, a little mathematics has been a boon. Too big a dose of mathematics has been detrimental. In some cases it has virtually destroyed the scientific content of the discipline and turned it into an abstract, academic game.

With poets and others who tend to use language without rhyme or reason, there is a kind of intoxication with language. This sometimes has produced dramatic results in poetry. However, it is a kind of madness which we cannot afford when human lives are at stake. In the long run, the contribution of mathematicians to *living* sciences seems to me rather paltry. In physics, they mainly have created beautiful fossils out of a dead science. However, although mathematicians may not really have much to contribute, they often make a great fuss over what little they *do* contribute. This often awes nonmathematicians into believing that they can contribute the most important ingredient of all to a science--infallibility.

In the future, there will probably be even more use of mathematical models. There will be even more pressure to use them in ways which are detrimental to the public and the public health. So it is rather important for the public to get a clearer idea of what mathematical language can and cannot do in a health issue. Otherwise, mathematics becomes a species of black magic that can be used to "justify" very dangerous deployments of technology.

WHAT IS MATHEMATICS GOOD FOR?

What can mathematics do? The first and foremost point is this: Mathematical languages are used in the sciences in the same way that any other technical language is used, as a way of describing events in the world about us.

What are the advantages of mathematical languages? There are three principal advantages. The first is that in mathematical language it is possible to make precise and unambiguous statements. This is often difficult in verbal languages. However, this is not an *inherent* virtue of algebraic languages. It mainly results from the fact that mathematicians do not know what they are talking about. Since the abstract symbols are meaningless in their own right, the symbols have to be explicitly specified in each model. In this process of specification, the meaning of the symbol can often be clarified and limited. In this way it is made more precise, sometimes overprecise.

The second principal advantage of mathematical languages is that the admissible operations in these languages are clearly defined. In a technical sense, they are *computable*. While this technical sense is somewhat peculiar (not quite what you might think from the ordinary sense), it is equivalent to saying that in principle the operation can be carried out on an electronic computer. The chief benefit of this computability is that linguistic manipulations in mathematical languages can be carried out and checked more rigorously than the manipulations of ordinary language.

This is, however, only a matter of degree. Modern linguistics theory has recently established that the natural languages such as English are also computable. However, it is still not practical to compute the linguistic manipulations of English except for special bodies of data (such as the surgical, operative reports we have been studying for 10 years).

The third principal advantage of mathematical languages is related to the first and second: When a theory is expressed in mathematical language it can be put to a more rigorous and objective test by the Galilean rule. In other words, if the theory and the facts can be both reduced to numbers, the assessment of the goodness-of-fit can likewise be reduced to a numerical operation. In principle this makes it possible to test and reject a defective mathematical model more easily. Very often theories that are stated in words have a good bit of "elasticity" and can be stretched to cover almost any facts.

This is *not* an advantage of a theory. Rather it is a serious disadvantage because then the theory will be nearly useless for prediction. The virus-cancer theory of the "oncogene" is an example of a very elastic verbal theory. It can "explain" almost any experimental result. Therefore, it really explains nothing. One contribution of the philosophy of science has been the notion that *a theory must be refutable in order to be meaningful*. This is a good point to remember.

The advantages of the mathematical languages are important if one goes by the Galilean rule and wants the operations going on in the black box to be objective and rigorous. Unfortunately, all these advantages are lost if the Galilean rule is ignored. As has been noted previously, for many mathematical models nowadays, no effort is made to abide by the Galilean rule.

When this happens, the original advantages turn into disadvantages. Precision of statement is lost when the tie to reality provided by the Galilean rule is severed. The symbols become unverifiable, and therefore meaningless. In their pure form, mathematical languages are, technically speaking, completely informationless, and consequently can tell us nothing about the world. The computability is of little value since it can now lead to statements that are false or meaningless. There is no longer any assurance that the "if" in the "if-then" statement is correct. Modern logic shows that if a false statement is put into the computations, then the output can be anything--true or false. Finally, the potential advantage of rigorous testing is lost if no test is made.

Consequently, an impressive, mathematical facade may conceal the fact that the computations are worthless from a scientific standpoint. Because of this potential for deception--a potential that has often been realized in such charades as the Rasmussen Report--the use of mathematical models may be literally worse than useless. Untested or untestable mathematical models can be hazardous to your health and safety.

OPERATIONS OF FIT

Let us now go back to the scientific argument at the NRC meeting in order to see more clearly what is needed to get useful results in mathematical languages. This brings us back to the Galilean rule and the question of goodness-of-fit between a theory and the facts. The schematic for the operations involved in testing whether a theory fits the facts is similar to the other schematics:

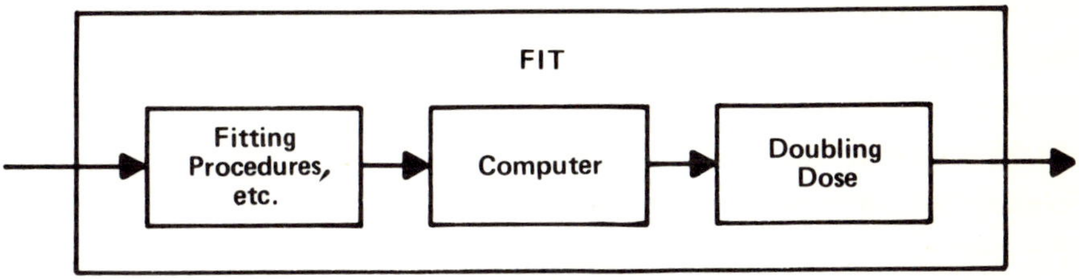

Fit for black box.

Once again the operations in the middle can be computerized, and therefore we are dealing again with computable linguistic operations. This has been the case for all three components of the original black box, and it follows that the technical or tactical operations involved in a scientific argument can be reduced to push-button operations. Thus, this phase of the argument can be checked out mechanically.

When a phase of the argument is strictly computable in this sense, there is not much point in trying to understand the mechanical details. Note that when the purely mathematical operations are computable they can be validated without having to know mathematics. This can be done without knowing how to read or write mathematical languages. If there is anything wrong here, it can be found by technicians. Scientific issues are not really involved in this part of the process. There is no reason for "expert opinion" to be involved in the evaluation because there is no room for "opinion" when a process is computable. It works or it doesn't.

DEALING WITH EXTRANEOUS FACTORS

The evaluation of the goodness-of-fit of a theory to the facts is generally not one step, but a process. This process may cycle and recycle if it is not stopped somewhere along the way by a fatal failure of the theory to fit the facts. The distinction between a terminal failure that stops the process (and rejects the theory) and a nonfatal failure is that for the latter, a further reconciliation of theory and fact may be possible. For example, there are three key parameters in the substantive model. If a poor choice of a numerical value for a parameter is made, this may produce a bad fit and a nonfatal failure. Some other values of the parameters may remedy this problem. Therefore, we can try another cycle in the fitting process rather than to reject the entire model. Note, however, that the specific model involving the poor choice of parameters is clearly rejected.

The step-by-step process here goes through several stages. Continuation to the next stage depends on success at the previous stage. This is a part of the continuous self-validating operations that are built into the mechanical structure of a sound scientific argument. Here, for instance, the first stage of testing the model is exemplified by the comparison of the observations and expectations in Figs. 5 and 6. Unless the fit is satisfactory, the process would stop here. Once the fit is found to be satisfactory by the usual criteria for goodness-of-fit, then the estimation of the doubling dose can proceed.

The dosage response curve in Fig. 3 is a good example of a second cycle of self-validation before the final validation cycle with the doubling dose. The mathematical model involving the first two components, the sampling model and the substantive radiation effects model does not deal directly with the dosage response relationship. The parameters of the model allow the proportion affected by a given dosage of radiation to vary with the dosage, but there is no restriction imposed on the way in which it varies.

If there were no effect from radiation, the proportion affected should be zero. This is clearly not the case. Conceivably, if there were some hypothetical bias between report of x-ray exposure and some other factor (e.g., age or socioeconomic status or what have you), then there might be a nonzero estimate of the proportion affected. *However, it would not show a consistent pattern.*

Note that there are other ways to rule out extraneous factors such as socioeconomic effects. The previous studies of the Tri-State Survey have looked at these factors. The relationships that were found are too weak to produce substantial biases.

This point was shown mathematically in two of my papers in the *Journal of Chronic Diseases.* One is titled "Spurious Effects From an Extraneous Variable" and is in the April 1965 issue. The sequel, titled "Pertinency of an Extraneous Variable," is in the June 1967 issue. What these papers show is that if there is a relationship between an extraneous stimulus variable and a response variable, the bias that can be produced in the study of another variable is an "echo" effect. In other words, it must be much weaker than the original relationships. Relationships which are weak to begin with cannot produce much of a bias. This is useful to know since it is always possible to dream up extraneous factors in these studies. It is useful to have ways of ruling them out individually or en masse.

The fact that a clear dosage response relationship appears in Figs. 2 and 3 makes it extremely unlikely that any factor other than radiation is producing this dosage response curve. While a plausible-sounding *verbal* counterhypothesis can always be given, for any factor to produce this strong a relationship as an echo effect would require a factor orders of magnitude *stronger* than radiation effects. This would make the hypothetical factor easily detectable, and there should be factual evidence on such a powerful factor. Apart from age and sex (which have been taken into account), no other factor nearly as strong as radiation has been reported.

The use of algebraic language here makes this point very clear. However, it can be confused in verbal language where the quantitative relationships are less clear. In this instance, mathematical languages have been used to good advantage, to clarify a complex situation.

In the face of a clear-cut dosage response relationship such as appears in these slides, casual arguments to extraneous factors simply show that the person doing the arguing does not know or understand the elements of epidemiology and statistics. These arguments don't discredit the data and findings; they discredit the person who uses these arguments. Indeed, when this kind of incompetent argument surfaces in a public health issue, it provides a good way for the public to distinguish its friends from its foes.

DOSAGE RESPONSE

To bring in the doubling dose, a third component of the mathematical model has to be used. This was previously labeled "dosage response." It is a separate little mathematical model entirely independent of the sampling and the substantive model. It involves two simple ideas. The first is the relationship between the definition of a "doubling dose for leukemia" and the three key parameters of the previous models.

Let D denote the average dose (in rads) for a given subseries. When all of the age and dosage categories for the men are considered, there are 13 subseries. Let D_2 denote the doubling dose. This is the unknown constant or parameter that is to be estimated.

By the definition of the doubling dose, the relationship between the percentage excess leukemia risk, PE, the dosage in a subseries D, and the doubling dose, D_2, can be written as:

$$PE = 100 \frac{D}{D_2}$$

Note that if $D = D_2$, then the percentage excess is 100 percent as it should be.

In "An Algebraic Argument" in Chapter 7, it was shown that

$$PE = A \times PI = A \times 100(R_L - 1) = K \times A$$

Equating the two expressions for PE, we obtain the simple algebraic relationship between the doubling dose, the average dose in a subseries, and the parameters of the mathematical model, A and R_L:

$$\frac{D}{D_2} = A(R_L - 1)$$

Note that the product term again dominates the expressions if R_L is large. Note also that the proportion affected is simply proportional to the dosage if D_2 and R_L are constant.

This formula does not involve any assumption about the world. It is a purely linguistic operation of "definition." A new symbol is brought in which is useful for a particular purpose at hand. It is defined in terms of symbols that were previously introduced. Nothing but a simple manipulation of language is involved. However, the introduction of a well-defined new term often precipitates a lot of unnecessary controversy and confusion. Indeed, arguing about "definitions" is a favorite ploy for tying a committee meeting up in knots.

The second idea is equally simple. It is the assumption that within the 1-rad range being considered here, a straight line will provide an adequate approximation. To avoid some unnecessary confusion, let me stress that *this* linear assumption involves *interpolation* and is a "weak" assumption. It should not be confused with the linear assumption used for *extrapolation*. That is a "strong" assumption and one that will often be seriously misleading.

The point here is one that anybody with some experience in curve fitting ought to know: A straight line usually gives passable approximation to a smooth curve over a short enough range. However, it can become a very bad approximation when extended over a wide range.

For the 1-rad range which is the focus here, the dosage response curve would tend to be gradually rising without too many abrupt twists and turns (small ups or downs don't matter). Hence, we know from experience that a straight line can give reasonably accurate estimates of the response at various dosage levels for this limited range.

When this dosage response component is added to the mathematical model, it makes it possible to directly estimate the doubling dose from the Tri-State Survey data. However the fitting procedure must be extended to do this job. The way this is accomplished by a computerized biostatistical technology is suggested by the schematic below:

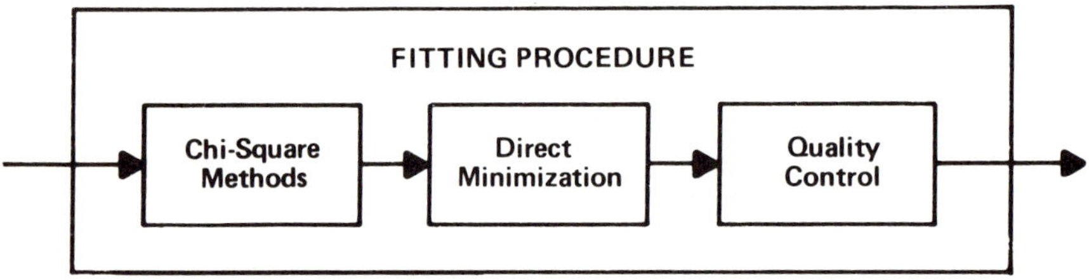

Fitting procedure.

CHAINS OF MODELS

One of the features of mathematical models that the reader may have already noticed is the possibility of hooking them together in sequence so that the output of one is the input of another. This has the potential advantage of building up to much more comprehensive and complex characterizations of the world than could be achieved starting de novo each time. Indeed, by proceeding in this way it becomes possible, in Newton's words, to "stand on the shoulders of giants." Even someone who is not a giant may then be able to see further than a giant.

Over a period of well over a decade, Dr. Blumenson and I have been modeling various stages in the overall process of cancer. They start at the point where radiation produces invisible damage to the structure of the DNA in the genetic material. They extend to the final stages of the process that have finally produced a free-living cell capable of going almost anywhere in the "salt water ocean" in the human body. The models cover (seperately) almost all of the intervening steps in this long and complex process involved in human cancer. My dream has been to chain all of these (suitably modified) separate mathematical models together to obtain a really comprehensive description of what human cancer is all about. Alas, this is likely to remain only a dream because of the inherent difficulties of this project. However, even if the interlinking is only hypothetical, the better understanding of one stage in the process suggests possibilities in the other stages.

Hence, with mathematical models it is possible to go in both directions, analysis and synthesis. The complexities of a given model can be avoided by analysis into submodels and on down to the elementary computer operations. Even without going all of the way, it is generally possible to analyze a complex model down to sufficiently simple and familiar components that the checking of these components is fairly straightforward. This means that most of the time the validation of a mathematical model in a formal or technical mathematical sense can be carried out systematically and in a largely mechanical way. This look at internal consistency of a theory is a bonus which comes from the use of mathematical languages.

Going in the opposite direction, synthesis, the actual construction of a complex model, largely consists of putting together simpler components. Most of the pieces can be "ordered" from a kind of mail-order catalogue. This "parts list" is provided by the mathematical or statistical or methodological literatures. Making all the pieces fit together and interact properly is not so easy. We also must check back every now and then to get a Galilean test of components or subassemblies. Finally, the test of the whole model is the hardest part of all. Not too many people have the talent to do this kind of work, and at present most of these people are being wasted on nonproductive activities in weapons technology and the physical sciences.

Perhaps this gives some idea as to what mathematical modeling *might* have contributed to the control of our major public health hazards. The trivial or neo-Jesuit use of the models is a sad misuse of this technological tool. What is even sadder is the thought of what might have been done with this tool--but wasn't.

COMPARISON OF STANDARD STATISTICAL METHODS FOR FITTING BIOMEDICAL DATA

Minimization of the chi-square goodness-of-fit statistic is plausible since the minimizing values of the parameters have an intuitively evident interpretation. They are the values where the theory best fits the facts. Other estimation procedures are used in statistics, among them "maximum likelihood" and "least squares," and there are partisans of still other methods. The choice is a matter of great concern to academic statisticians. The statistical literature is filled with papers on this topic. The truth of the matter is that there are some good statistical methods (among them, minimum chi-square) and some bad statistical methods. It is important to choose a good method. However, it generally makes little difference in practice which one of the good methods are used.

Why then the great debate about "optimum" techniques? This is simply another example of the tendency of the academicians to focus on pleasant theoretic trivia and then ignore the difficult practical issues that are important. From the standpoint of publication and academic promotion ("publish or perish"), the payoff is for trivia. As can be seen from the literature of statistics and other disciplines, trivia is easily publishable. Important work is rare. It is very hard to publish.

Now, if academicians want to waste their time, this is their prerogative. However, they shouldn't intervene in public health issues and mess them up with this kind of trivia.

One thing that computers have done is to make it relatively easy to do the calculations for two or three or half a dozen different statistical methods. What should have been learned from the "horse races" between rival methods, a favorite sport of academicians, is that there has been much ado about nothing. The good methods come together in a dead heat. This is only to be expected. They differ only in secondary or tertiary approximation to the theoretic optimum. This has been shown formally in a mathematical appendix to "Statistical Testing of a Deep Mathematical Model for Human Breast Cancer" in the *Journal of Chronic Diseases* in 1968. Similar results have been derived by others. But as long as it is much easier to publish the results of horse races than to publish substantial work with new ideas, trivia will be the name of the academic game.

When a public health issue gets bogged down in statistical technicalities, the public should realize that the disagreement between experts probably has very little to do with the actual issue.

The controversies between the proponents of various statistical methods of estimation date back to the days when calculations were made by hand. Since all of the fitting methods involve finding a minimum (or a maximum), a little finesse in this procedure could save days of tedious computation at the turn of the twentieth century. Hence, in those days this fussing around might have made some sense. Nowadays computers are so fast and so powerful that minimization can be done directly by brute force. Once again, it doesn't really matter *how* the job is done. The end results are the same.

Personally, I prefer brute force. In this way I can see every step of the way to the minumum. Sometimes, as has happened to me, a programming or other artifact leads to a false minimum. With graphical methods this can be seen directly in time to correct the result. This inelegance and "waste" of computer time seems to shock my colleagues. However, it is the quickest, safest, and cheapest way to an answer in complex minimization processes.

Perhaps the main point is that the choice of method is a matter of personal taste which has nothing much to do with the health issue. Technical discussions of minimization procedures hold great appeal to the experts. However, I suspect they are simply a way to bolster the egos of middle-aged and deteriorating males.

JUDGE FOR YOURSELF

What is important in the scientific argument? What should you try to understand in order to come to your own decision on a health issue? It is important to try to get a broad picture of what is going on in the data, in the analysis, and in the final conclusions.

Here, for instance, we started with a complicated contingency table for radiation, leukemia, heart disease, and age. The first step in the analysis was to set up a mathematical model which would explain this data--which would give reasonably good predictions of the numbers in the table. That is the business of theory. That was done.

The next step was to look for patterns in the data such as are represented here by the dosage response curves. A scientific argument requires us to establish that the patterns we find cannot be explained away by sampling variation (that they are not due to chance). Confidence intervals are a tool for establishing this point. The pattern here shows that increasing doses of radiation in the 1-rad range increase the risks of disease.

Having established the patterns, the next step is to use this new information to answer the specific questions of interest. Since the doubling dose can be calculated directly from the parameters of the model, this leaves only the not-so-simple technical task of finding the value of the doubling dose that minimizes the chi-square. But the details of doing the job is not something that the general public has to understand. Mistakes in such purely computational procedures can be easily found and will be widely publicized.

By taking the lid off the black box I have tried to dispel the aura of mystery and magic that only obscures genuine scientific research. This mystique is part of the cult of sciencism and the worship of technology. It has nothing to do with science. The same kind of mystique occurs in medicine, law, and other professions. It is merely a way to shroud professional activities from the eyes of the public. It has been used in all priesthoods for the same purpose. The mystique is popular with those who see themselves as consecrated high priests of technology. Taking the lid off the black box shows that there are only mechanical operations (linguistic transformations) going on inside. This shows that the mystique is self-serving nonsense.

To avoid leaving the reader with the impression that taking the lid off the black box serves only to refute the current scientific mystique, let me conclude with a few positive points. As has been seen over and over again, the whole point of the sometimes elaborate machinery of the scientific argument is to get to the truth--to get reliable information about what goes on in the world about us. In scientific language as in ordinary language, honest reporting and common sense indicates the general direction to go. The main distinction between a scientific and ordinary argument is one of degree--the scientific argument has complex safeguards against error.

Statistical technology provides a tight quality control, but it does not eliminate error entirely. When properly used, it insures that there is only a very small probability of error--that errors will be very infrequent. In this imperfect human world, nothing can provide more assurance of the truth than this. The quality control, as the diagrams show, occurs at every link in the chain.

In a good study, the scientific argument is self-validating. If, for instance, the dosimetry is challenged, we can look at the dosage response curve. What many scientists (indeed, many epidemiologists and statisticians) don't seem to realize is that if a measurement or classification is not very good, the effect will be to *destroy* the patterns that might be in the data.

Years ago I started what is nowadays a whole literature on misclassifications with a single paper entitled "Misclassification in 2 × 2 Tables." It appeared in *Biometrics* in December 1954. Since then there have been dozens of papers that have confirmed and extended the original findings. Only under very special conditions can misclassifications in 2 × 2 tables put patterns into the data. Any adequate study avoids these conditions. So when we find a clear pattern in the dosage response curve, this answers the objection directly from the data.

My main point is that a good scientific argument with quality controls and self-validation tells the whole story, and tells it objectively and honestly. There is no need for the intrusion of biased, subjective, and often erroneous "expert opinion." So when "expert opinion" becomes the basis for the review of a specific paper, a decision whether a research program will be supported, or in other decision-making in science, this can only be counter-productive. It puts back in the biases and preconceptions and conventional wisdom that a well-designed study of human problems screens out. The much-admired peer review procedures violate the basic principles of modern sciences.

Worse yet, the intrusion of "expert opinion" into decision-making deprives the public of the benefits that it could get from good scientific research. In this specific study, the NRC should have taken immediate action to reduce the permissible levels of radiation to 0.5 rads. No other carcinogen is permitted at anywhere near a dose that is known to substantially increase the risk of cancer in humans. This Tri-State data has sent a clear message that could protect the public. However, the NRC disregarded this message and relied on a chorus of "expert opinion" to drown it out. This intrusion of expert opinion is not only the opposite of genuine science; it is also very much against the public interest.

FINAL EXAM

Now that you have come through a cram course on the nature and structure of a genuine scientific argument on public health issues, you might like to test your grasp of these ideas by criticizing a recent federal interagency report on low-level ionizing radiation hazards (Biological Effects of Ionizing Radiation, Science Work Group, 1979). Produced by a collaboration of almost all of the federal health and science agencies, the report concludes that the "results are not conclusive" and "further epidemiologic and experimental work is needed to clarify the possible health effects of low-dose radiation." The "conclusions" are, of course, asserted "in the name of science." What do you think of the following arguments?

1. The report cites a series of more than half a dozen recent studies of populations exposed to low-level nuclear radiation. These include studies of servicemen exposed to

fallout at the Big Smoky weapons tests and the Utah children downwind from these tests; studies of occupation exposures at the Hanford reprocessing plant in the state of Washington and the Portsmouth Naval Shipyard in Maine; and studies of exposures to nuclear wastes such as uranium tailings or radioactive emissions. In each case, the facts show an excess of leukemia or other cancers. However, the members of the radiation protection community who wrote this report give each independent scientific study a negative "interpretation." It is then concluded that the evidence is "inconclusive." Is this science?

2. The report reviews another half a dozen recent studies of diagnostic radiation. Again the facts show an excess of leukemia in every study. Again each study is given a negative "interpretation." Does this warrant a conclusion that the results are "inconclusive"? What are the chances of a dozen "false alarms"?

3. The estimate of doubling dose at 5 rads is criticized as lacking "independent confirmation from other data sources." All of the above-mentioned studies (where the risk of leukemia is two to six times greater than expected) involve exposures in the 1-rad range. Hence, a 5-rad doubling dose is consistent with the results of these studies. On the other hand, if the doubling dose were 50 to 100 rads (as the report claims), then no such positive reports would be expected. Is the assertion about confirmation based on fact or "interpretation"?

4. A list of research on ionizing radiation supported by federal agencies such as the Department of Energy and the National Cancer Institute is included in the report. Apart from the support of our studies and the work of Mancuso, Stewart, and Kneale, it is noteworthy that none of the positive work was supported by Big Science agencies. The work that had been supported was cut off when it gave positive results. On the other hand, none of the work supported by these agencies was positive. Indeed, very little of this work has been of any

value in protecting the public health and safety. What conclusions can you draw from this about the "research" of the Big Science agencies?

5. (Homework question.) What can you do about the federal agencies that you are supporting as a taxpayer when they collectively act against your interests (as in this report)? What can you do when the agencies (a) insist on lying about radiation hazards, even though the public is now well aware of the truth; (b) refuse to take any effective action to save your life from the technogenic (techno = technology, genic = caused by) diseases that are now our number one public health problem; (c) do not feel accountability to the public or its representatives and consistently act in an irresponsible and unresponsive way? Do you really need such agencies, or would you be better off without them?

chapter 11

POLITICS AND SCIENCE

The role of politics in the implementation of public health strategies.

THE RELEVANCE OF POLITICS

The first part of this book (Chapters 2-6) introduced the basic scientific strategies for studying health hazards in the environment and used the hazards of diagnostic x-rays as a continuing example. The next part (Chapters 7-10) considered in more detail the application of these strategies to the task of drawing reliable conclusions about health hazards and their control, again using radiation-induced iatrogenic diseases for the continuing example. Informally speaking, one could say that the emphasis has been on the "science" rather than the "politics" of public health although, as in any actual health issue, the "politics" comes up from time to time. In this final part of the book, the emphasis will shift to politics (though science will come up from time to time). In practice, the two topics cannot be hermetically sealed off from one another.

Some readers may consider the discussion of the politics of science inappropriate or distasteful. Other readers may find this topic more interesting than the sometimes tedious task of constructing a valid scientific argument. Each reader can decide whether or not to go on to this final part. However, perhaps it is worth briefly considering at this point whether or not it is appropriate to mix science and politics in a book on scientific strategies for saving lives.

For some scientists, "politics" is a dirty word. When I was young and innocent I felt this way too. Just talking about politics seems to pollute the pure image of science cherished by some idealists. However, I make no apologies for bringing politics into a discussion of scientific strategies. It would hardly be "science" to ignore what is clearly a major factor in the action situation in most public health issues. The "politics" will be there whether it is talked about or not.

Some of my academic colleagues reproach me as an "activist" because of my view of the job of an epidemiologist. I don't think my job is to get grants or to write papers or to give lectures. *The job of an epidemiologist is to stop epidemics*. The only way to do this is to go out into the turbulent real world about us to eliminate or control the health hazards. With the technogenic hazards of our society this means somebody's ox will be gored. There is likely to be a bitter battle to protect the vested interests that are threatened. Inevitably there will be politics. If the stakes are high enough, there will be dirty politics. So whether one is a health professional or a concerned citizen, some understanding of the political game is needed if effective action is to be taken to save lives.

What can be said about the politics of public health that will actually be useful in implementing the life-saving strategies discussed previously? This is a hard question to answer. I can only hope that this and the next four chapters will be useful. The rest of this chapter will briefly consider the political and scientific repercussions of the data on diagnostic x-rays that was presented to the NRC. Chapter 12 gives some guidelines on an aspect of political action that is sometimes overlooked, the role of language in public affairs. The guidelines were written for scientists who want to tell the truth in Washington. However, they can also be used by concerned citizens who wonder whether a scientist is telling the truth or not.

Chapter 13 deals with risk-benefit analysis. In principle, this approach could be used to resolve health issues in a scientific manner. In practice, however, risk-benefit analysis has generally been a gimmick for lobbyists. Chapter 14 goes into the very dirty politics of the low-level radiation issue. It is based on a 1978 congressional hearing which brought to light the way that the Department of Energy and the National Cancer Institute tried to suppress the new findings on radiation hazards. Finally (as something of an antidote to Chapter 14), Chapter 15 briefly considers how the scientific strategies in this book have been instrumental in actually saving hundreds of thousands of lives despite the last-ditch political opposition to protective measures. Of course several times this number of lives could have been saved if the measures had not been delayed or blocked by the lobbyists for hazardous technologies.

CONFIRMATION OF THE DOUBLING DOSE

Many months have passed between the time that the presentation was made to the NRC and the present, and in this time too much has happened on the issue of low-level radiation to detail here. There has been some good news and some bad news.

The good news is that there has been some movement to protect the public against x-rays and other radiation hazards. This has largely been a result of public pressure rather than the result of any action by the NRC or other federal agencies. Three Mile Island and other media events have made the public very aware of the dangers, although the federal agencies continue to issue official reports denying any serious hazard from low-level radiation. According to public opinion polls, three out of four American women recognize the hazards of low-level radiation and are trying to protect themselves and their small children from these hazards. Although no polls of the scientific establishment are available, the ratio would probably be reversed [1]. All that such a poll would prove, however, is that there is now a ridiculous situation where housewives know more about radiation hazards than many health physicists, radiologists, and other so-called experts. The truth is out.

From a scientific standpoint there was also good news in a series of new studies that have confirmed the estimates of the doubling dose of leukemia reported to the NRC. My testimony of March 6, 1979, to the Senate Government Affairs Subcommittee on Energy, Nuclear Proliferation, and Federal Services in Washington, D.C. began by noting this point:

> Three years ago it was widely believed by the self-styled radiation protection community that it would be impossible to detect any health effects in studies of people exposed to dosages in the 1-rad range. At that time, Tom Mancuso and I were the only ones doing large-scale epidemiological studies to look at these hazards. Two years ago I predicted that if scientists would only try to look at populations with exposures in the 1-rad range they would find, as we did, that there are serious health hazards. Since that time more than half a dozen new studies have looked at what happened to persons exposed to nuclear radiation in the 1-rad range and have reported positive results. These are the studies that I want to try to put together.
>
> In 10 minutes I cannot hope to go into details on all the studies, the criticisms of these studies that have been made by the members of the radiation protection community who wrote the interagency report, or the answers to these criticisms. Very briefly, there are three kinds of studies of nuclear radiation hazards at the 1-rad level. The first kind deals with persons who were exposed to fallout from the nuclear weapons testing of the cold war era. This includes studies of the servicemen at Big Smoky and other tests. There are also the after-effects on adults and children in the areas of Utah downwind from the tests. The second kind of study involves occupational exposures. This includes studies of the workers at the Hanford reprocessing plant and at the Portsmouth Naval Shipyard. The third

> class of study involves exposures to nuclear wastes such as the uranium tailings or releases from power plants. Dependinq on what is counted, there are now between half a dozen and a dozen positive reports of hazards to persons exposed to nuclear radiation in the 1-rad range. It is virtually impossible that they are all false alarms.

This testimony involved an early draft of the Interagency Report, commonly called the Libassi Report, but the bibliography of the final version [2] will be used here. Numbers in brackets refer to references at the end of the chapter.

More specifically, the final version of the Libassi Report cites five references for the hazards of nuclear radiation from fallout if thyroid cancer is also considered [4-8]. However, this list only includes publications in the technical literature. It omits the reports on fallout from Dr. Ernest Sternglass and others, even when they appear in a congressional report [9]. It omits media reports entirely, for instance the reports on the marines at Nagasaki [10]. The coverage of hazards to workers at nuclear installations is better, and seven references with positive results are cited [11-16]. There are three positive reports on hazards of nuclear wastes or emissions or areas of high natural radiation [17-19], but none of the studies of populations in the vicinity of nuclear power plants [20] are cited. The Rocky Flats and uranium tailings hazards are given only cursory mention. Despite the omissions, it can be seen that there are well over a dozen positive studies which were cited in the Libassi Report, disparaged, and then disregarded.

There are 11 reports of positive findings for diagnostic x-rays cited [21-31], all of which find excess leukemia among patients exposed to this low-level radiation. A negative study of occupational hazards of radiologists is cited [32], but not the positive studies. An important study of the *children* of radiologists [33] is omitted, as are some important diagnostic x-ray studies [34].

One might wonder why in 1980 there are so many positive studies on groups exposed to low-level radiation when in 1960 or 1970 there were so few. Basically, what has happened is this: *Time is running out on both the threshold hypothesis and the linear hypothesis.* The nuclear exposures started in the 1950s and 1960s, but due to the long latent period for the malignant diseases the health effects are only now coming to light.

HOW POLITICS COME IN

When the scientific reports documented in the previous section are considered together with the Tri-State Survey findings presented earlier in this book, there can be little scientific question that low-level radiation has very serious health hazards. The only way to "answer" such overwhelming scientific evidence is to turn the scientific argument into a political one which masquerades as "scientific." This was the political strategy of the tobacco industry many years ago when we first presented the scientific evidence on the health hazards of cigarettes. The evidence on the health hazards of low-level ionizing radiation is just as strong as the cigarette-cancer evidence, and the political tactics used to deny the epidemiological evidence are similar in both issues.

Even today the lobbyists for the tobacco industry try to pose as scientists and to write "critiques" that deny the health hazards of cigarettes "in the name of science." However, nowadays few people believe these lobbyists. The radiation lobbyists also pose as scientists. They too write "critiques" to deny the hazards of low-level radiation. In both cases, the so-called "scientific critiques" involve very little science. They are simply a device for political lobbying.

However, there is one major difference between the lobbyists for the cigarette companies and those for radiation technology. For cigarettes, the National Cancer Institute and the industry lobbyists are on opposite sides. For radiation, the lobbyists are entrenched in the National Cancer Institute, the National Academy of Sciences, and other federal agencies.

At this point, some readers who have followed the scientific strategies presented earlier may have some difficulty in understanding how, in the face of the factual evidence now available, it is possible for persons who are supposed to be scientists to continue to deny the serious health hazards of low-level ionizing radiation. The previous section cited a series of independent reports on populations exposed to average doses of radiation less than 5 rems where the data show double or triple the normal risk of leukemia. These reports are compatible with a doubling dose under 5 rems and the worst possible case (curve C) in Fig. 1. They are incompatible with the doubling dose of over 100 rems claimed in the Federal Interagency Task Force Report and the linear hypothesis used to get this extrapolative estimate. They are also incompatible with the infinite doubling dose of the threshold hypothesis. So of the three hypotheses in Fig. 1, it is unfortunately the hypothesis that is least favorable from a public health standpoint that fits the facts.

However, neither science nor logic are involved in the official federal reports that still insist that low-level radiation is harmless or nearly so. For example, the facts on the Big Smoky bomb tests *reported by federal agencies* flatly contradict these claims. The Defense Nuclear Agency carried out a retrospective reconstruction of the troop exercises at Big Smoky and estimated the mean exposure at 970 millirems--less than 1 rem. The Center for Disease Control found "about double the statistical expectation" in its leukemia data. Even the Interagency Report admits "an expected incidence of between 2 and 4 cases" of leukemia from the radiation. This would virtually be impossible statistically with a doubling dose over 100 rads or an infinite doubling dose. There is no way to reconcile the facts in the Interagency Report with the conclusions of that report.

Where, then, did the conclusions come from? They didn't come from science, they came from politics. The Interagency Report seems to have been specifically written to enable the Veterans Administration to use it to deny compensation claims from the servicemen at Big Smoky and other nuclear tests. According to the National Association of Atomic Veterans, this is certainly how the report is now being used.

The stakes here are high: These compensation claims from servicemen, civilians exposed to fallout, nuclear workers, and others exposed to low-level radiation could run into the billions of dollars. Hence, the myth that low-level radiation is "harmless" (which was originally created to protect the nuclear weapons testing programs and was maintained to protect the nuclear power program) has now been revived in order to balance the federal budget at the expense of the servicemen and others exposed to this "harmless" radiation. This is politics at its worst.

The reason for bringing politics into a discussion of scientific strategies for health issues is that there is no way of keeping politics *out* of these issues.

COPING WITH POLITICS

There can be no magic formula for coping with politics since every individual has his or her own chemistry. Each of us has to find ways of coping with political pressures that will work for us. Generally this takes trial and error. The most that can be done here is to suggest a few tactical tips for coping with politics. These reflect lessons painfully learned in 30 years of public health politics.

The first tip is simply this: Know what you are getting into when you get into politics. It is a little easier to cope if you know what to expect. In public health politics, expect the worst.

Whether you are dealing with radiation hazards or some other side effects of the misuse or abuse of technology, there are likely to be large amounts of money involved. Usually the higher the stakes, the dirtier the politics. So if you are invited to a hearing or debate or meeting, don't suppose that you are invited to a tea party. This isn't the relatively civilized politics of the academic world, nasty as this politics can sometimes be. This is jungle politics where the strategy is simply: Go for the jugular vein. At a "debate" which is supposed to discuss new research findings, the whole point of lobbyists will be to discredit or kill the findings in any way possible.

Any scientist whose findings are seen as a threat to governmental or commercial interests or programs can expect political retaliation in the form of what are politely called "disincentives." My 1978 congressional testimony described three of these:

> Let me now turn briefly to the second topic, the disincentives for research with positive findings on low-level radiation hazards. There are three main disincentives. The first is suppression of the findings by blocking publication or presentation at meetings. The second is attacking the reports and their authors by the professional version of what is ordinarily called scurrilous gossip. The third disincentive is to cut off the funding of the research. All three disincentives use peer review as the instrument. By peer review, I mean only the small, closed-in groups that control much of the decision-making in the sciences. There are a lot of varieties of peer review.
>
> The other side of the coin are the incentives given for negative findings. These include publication, appointment to prestigious committees, increased fundings, and so forth. This structure of incentives and disincentives is one of the main reasons why the public has received so little of value for the large sum of money spent on radiation research.
>
> We, in a sense, are rewarding expertise in not finding anything positive, and so we get that kind of work.

Knowing what the disincentives are does not make them any less painful, but it may help to avoid the natural reaction of shock or surprise. Innocents from the academic world or well-meaning concerned citizens may find their first encounter with the jungle politics of these health issues to be a shattering experience. What may be especially discouraging is the realization that the federal agencies that one would naturally look to for help are likely to be far more concerned with protecting their constituencies than with protecting the public.

Not only is it important to go into the political arena with one's eyes open, it is important so speak plainly and bluntly about what one sees there, even if it may seem impolite to do so. My second piece of advice is this: You are not going to save lives--including your own--if you act like perfect little ladies and gentlemen. When in the jungle, you have to learn a little jungle fighting. Your adversaries won't be "nice guys."

My 1978 testimony gave this example of how the second disincentive works in practice:

> When a Jack Anderson column reported our fate and that of others who had positive findings--a copy of the column is submitted--and Dr. Bernard Cohen wrote a letter in which he lists a series of agencies which he claims have reviewed and rejected our work.
>
> Dr. Cohen has been associated with the Atomic Industrial Forum, which is the industry lobby, and his libelous statement is simply a lie. None of the agencies have reviewed our work, in a formal sense, nor rejected it formally. In fact, the work is coming up for its first formal review when I will present it at the Nuclear Regulatory Commission hearing on April 7, which is a meeting held specifically for this purpose.
>
> Attacks by gossip and innuendo on scientists who come up with positive findings have hurt their careers, and there is really no defense to this kind of thing. And I would like to add one more example of this that happened in my Friday testimony, because Dr. Victor Bond, speaking for the National Academy of Sciences, stated at the meeting on February 10 on this topic, you know, the same thing. I was reporting our doubling dose findings; he said this work has been reviewed and rejected. It has not even been seen.
>
> Mr. Rogers: Who was it who said that?

Dr. Bross: Dr. Victor Bond. He may have been confused by something, but what he said is wrong. It is a pretty damaging statement to make, this, for a scientist.

How can one cope with this kind of politics? Sometimes there is nothing better than to grin and bear it, hard as this may be. At the present time the lobbyists for radiation or other technologies assume that they have the right to discredit a scientist in any way they please. Ironically enough, if a scientist objects to this kind of abuse by lobbyists, the response is likely to be indignant cries that the objection is an attempt to limit "freedom of speech" or "open scientific debate"!

POLITICS AND COMMUNICATION

The disincentives applied to the scientists with positive findings of low-level radiation hazards can be viewed in different ways. Some ways may be more helpful than others in coping with these disincentives. An obvious way to regard these disincentives is as examples of the politics of repression. However, this view, while not unrealistic, may lead a health professional or concerned citizen into a defensive or even a hopeless attitude that might be called a "siege mentality." It is sometimes hard to avoid falling into this dismal attitude since it is discouraging that in what we like to think of as a free society there are federal agencies that operate in this manner. This contrast between the world as it should be and the world as it is will be especially painful for anyone with even a touch of idealism.

However, to cope with the politics of repression, it is important to understand how the machinery works. This is a first step toward putting a monkey wrench in the works. The third tactical tip is helpful in understanding the mechanics:

In politics, the name of the game is communication.

Looking at the disincentives in this light, it becomes obvious that the disincentives are designed to cut off a scientist from communication with the public or with other scientists. For most scientists, communication with peers is essential. Praise from one's peers is sometimes the greatest reward that a scientist can hope for. It is not really necessary to send a scientist off to physical exile in Siberia to have effective repression; an intellectual isolation from the scientific community is just as effective.

If the underlying strategy of the disincentives is to sever communication and to isolate a scientist, the obvious counterstrategy is to restore or improve communication. However, it isn't easy to achieve this goal. The channels of communication in the scientific community are, for example, publication in technical journals or invitations to scientific meetings. The first disincentive closes off these channels. This is relatively easy to do because most scientific organizations are controlled by tight little in-groups, and the rank and file have little to say about the operations. Although there is a great show of "independence" by the various organizations, the fact is that they are subsidized by the federal government in a variety of ways. For instance, "page charges" are paid by federal grants to cover publication costs, and without such subsidies some journals couldn't operate. So communications which might offend or antagonize the federal agencies are not very welcome. This kind of genteel censorship is more effective than a government order not to publish.

Consequently, it is hardly surprising that the material presented at the NRC hearing was hard to publish. Some of it appears for the first time in this book. Among the journals that refused to print this material was *Science,* where Dr. Philip Abelson, the editor, personally handled the review.

The machinery for the first disincentive may be better understood from a brief look at what went on with the part of the NRC material that was finally accepted by Dr. Alfred Yankauer, editor of the *American Journal of Public Health*. There were strings attached to the acceptance: I had to agree to simultaneous publication of a "critique" by Drs. John Boice and Charles Land, two staffers of the National Cancer Institute. While I knew what was coming, I thought

I would get a chance to reply to the attacks on our work. However, the editor scheduled the editorial processing in such a way as to preclude a response in the same issue or, indeed, anything more than a brief letter of response.

The extreme editorial bias in the publication process is shown by the Editor's Note which included a statement with evident similarity to those previously cited in my congressional testimony. In this note at the start of the article the editor said:

> Its subject matter is of great importance to the public health, yet Dr. Bross stands virtually alone in defense of his data and in the interpretation he places on them.

At the time this was written, the editor and the two NCI staffers he had picked for the derogatory critique were the only persons who had seen the article under review. The "jury of my peers" had not seen the evidence, and the "judge" was making this unfair attempt to sway the "jury."

The wide use of this particular "dirty trick" is testimony to its effectiveness in discrediting a scientist and his work, and this Editor's Note was widely used. It appeared in news stories, including those in technical journals. The only mention of our findings that has ever appeared in *Science* was accompanied by this "tag line." Whenever I talk about radiation hazards on television, the line is almost certain to be quoted against me, irrespective of the subject.

On the evening of the first reports on Three Mile Island, I was discussing on television the dangers of the 30 million curies of radioactivity stored as liquid waste in a deteriorating carbon steel tank at West Valley, New York, near Buffalo. My copanelist, a Westinghouse engineer, first reassured the audience that the news reports on Three Mile Island were much ado about nothing. Then she "answered" me by reading the Editor's Note. When I was talking about the danger of a proposed mammography screening program of the National Cancer Institute of Canada in Toronto in January 1980, the same thing happened. When the head of this NCIOC program wasn't doing so well in the debate, out came the Editor's Note. It seems to be the "answer" to any unpleasant facts.

After the contribution of the editor, the subsequent "critique" by Drs. Boice and Land was something of an anticlimax, although it was longer than the original paper. I briefly considered, but rejected, the possibility of including this "critique" in this book. It contains a rather complete catalogue of all of the gimmicks and "dirty tricks" that have been used to discredit the positive studies of low-level radiation hazards, but it would take too long to point them all out. There is, for instance, the gimmick mentioned in the NRC presentation of using a faulty statistical analysis to "refute" a sound one. The stock arguments against any epidemiological study are all here. The "critique" is a perfect example of its genre.

The perfection was born of long practice. Dr. Charles Land has "critiqued" a long series of studies for the National Cancer Institute, both before our report and after it. He had attacked Dr. Thomas Mancuso and his distinguished coauthors, Dr. Alice Stewart and George Kneale, repeatedly both verbally and in print. He has performed a similar service for our work on several occasions. He has attacked the reports of Drs. Thomas Najarian and Theodore Colton on hazards at the Portsmouth Naval Shipyard, again repeatedly. He has attacked the studies of the children in Utah and the veterans at Big Smoky. If there is any positive report that he has not attacked, then it has not come to my attention (or his).

Is there any good way to respond to "critiques" of this kind on public health issues? If there is, I don't know it. Perhaps the best that can be done is to point out the track record of the persons or groups that generate such "critiques" and to apply the following tactical rule:

Once is a critique. Twice is a hatchet job. Three times is a professional hatchet man.

When one person repeatedly attacks the work of other scientists this may be labeled "scientific criticism," but is really the hallmark of a lobbyist. In general, it is a waste of time and effort to try to talk science to a lobbyist. What the lobbyists who pose as scientists are most afraid of is being publicly identified as lobbyists. This destroys their credibility. This is likely to be more effective than trying to "answer" the false, misleading, or gobbledygook statements in these "critiques."

FEDERAL AGENCIES VS. THE PEOPLE

Once you get involved in the politics of public health, you never know where it will end. As an individual you must cope with organized groups--federal agencies, professional groups, industry lobbies, and environmentalists. Dealing with organized groups is often like navigating in a fog--you can never be sure what you are getting into, just where you are, or what direction to go. It is very confusing. However, you cannot wait until the fog clears because it never does. To get anywhere, you will have to make guesses, take chances, keep going, and hope for the best.

One word of advice in dealing with organizations: *Don't make the mistake of trusting in names, titles, or labels*. Don't suppose that what an organization ought to do by logic or common sense is what the organization will actually do. Remember: *The top priority of any agency or group is to perpetuate itself*. So pay no attention to promises, just look at performance.

If you don't expect much from the federal agencies that you are supporting as a taxpayer, you won't be so disappointed when you get very little from them. Don't expect service agencies to serve the public because they really serve their constituencies. The second priority of the federal agencies is to help those who can, in turn, help the agency. In practice, this means other organized groups, since individuals and an unorganized general public have very little clout. This may sound cynical, but in dealing with the federal agencies you almost have to be cynical.

Thus, if you go to the National Institutes of Health, don't expect them to be concerned about your health; their concern is the health industry. If you go to the Veteran's Administration, expect the rules to be tilted against the veteran. If you go to the Justice Department, don't expect justice. Don't go by the names.

A few personal experiences with how the system works in Washington may give you some idea of what you are up against. Let me start with a political sequel to the NRC presentation. In the summer of 1978, I received an invitation to go to Washington to meet with Mr. Peter Libassi, Chief Counsel for the Department of Health, Education, and Welfare under Secretary Joseph Califano. The nominal topic of the meeting was: What were the implications of our new findings for the compensation of the servicemen exposed to the nuclear weapons tests at Big Smoky and elsewhere?

This topic first came to public attention at the congressional hearings in January and February 1978 that have been previously cited in this chapter. These radiation hearings are an example of congressional hearings at their most effective and productive, largely because of the careful spadework of the co-chairmen, Congressman Paul Rogers of Florida (Democrat) and Tim Lee Carter of Kentucky (Republican), and their staffs. The testimony in Serial 95-179 is a gold mine of information and should be carefully studied by historians and political scientists as well as by concerned citizens.

For more than an hour the DHEW lawyers and Mr. Libassi discussed a memo that I had prepared on the difficult questions of compensation for the subtle and long-delayed health effects of the exposures to low-level radiation from fallout. Our work could give clear answers to some of these questions, but others could not be answered with any certainty. For instance, there is always a "Single Case Paradox"--we can be confident that *some* of the cases in a series are produced by the radiation exposure, but we cannot say *which* cases these are. The Primacy Principle applied in this situation, I argued, and the benefit of the doubt should go to the veterans and their widows. It is interesting that the *Atomic Veterans' Newsletter* for December 1979 says: "If the burden of proof is to be placed on the shoulders of the veterans and widows, then we will proceed to assist each other until justice prevails."

At the time, however, the discussion with the lawyers (including VA lawyers) had been so frank and free and in such substantial agreement that I was beginning to feel hopeful that for once a trip to Washington would accomplish something. This euphoria did not last very long. The next step was to discuss our findings with the representative of the Task Force who appeared to be operationally in charge of compiling the section on biological effects. It was Dr. Charles Land.

Although not on the best of terms with Dr. Land, I dealt with him as the official Task Force representative. Hence, it seemed appropriate to present a frank portrait of our research, "warts and all," and to answer in full all of his questions, however negative in tone they might be. After all, this was "privileged information." The session lasted much longer than the one with Mr. Libassi, and Dr. Land seemed to accept at least some of my answers to his criticisms. However, it was not entirely a surprise when the portrait of research in the published "critique" turned out to be only the "warts." None of my answers to the criticisms were even mentioned.

While it is natural for all of us to react to a situation in terms of its impact on us, whether favorable or unfavorable, for effective political action it is important to try to see the broader picture. In other words:

In politics, what happens to an individual is less important than what happens to a population.

Here, for instance, what worried me most about the presence of Dr. Land in this very sensitive position was not what it meant for me, but what it was likely to mean for the thousands of veterans who would be coming in to the VA for compensation. It looked bad for them.

The question was: Why had someone with Dr. Land's lengthy track record and special talents been chosen for the Task Force review of the new scientific evidence on low-level radiation? If a fair and objective review of the new positive studies had been wanted, it would have made sense to choose someone who was not already on record with attacks on each and every one of the studies under review. Was this choice by accident or design? This latter question is always a tricky one in Washington. It is always hard to rule out sheer ineptitude and incompetence as the reason for decisions of the federal agencies.

In these political affairs one is always in a foggy situation and never can be sure what is going on at a given time and place. Only in long retrospect do these events become a little clearer. Apparently I had run into a situation which was rather like finding a tobacco industry lobbyist writing the next Surgeon General's report on cigarette hazards. Yet I couldn't be certain what this signaled. Only when the report was out and the policy in effect could I be fairly sure what had gone on.

Now that the Interagency Report has been published and the VA is using it to deny compensation claims, it is fairly clear in retrospect that someone up there wanted a report that would minimize the compensation costs. That individual chose the right tool for the job. As a result of that high-level decision, the VA has continued to refuse service connection to most of the servicemen who are now sick or dying from leukemia, lymphoma, or other diseases induced by their low-level radiation exposures at Big Smoky and the other nuclear tests.

Another example of how the system works involves the third disincentive that was mentioned earlier, the cutoff of funds for research that turned up positive evidence of low-level radiation hazards. At the 1978 hearing, I testified how an angry radiologist from M.D. Anderson Hospital in Texas and two other members in good standing in the radiation protection community came to the Roswell Park Memorial Institute on the site visit that reviewed our study. They came there to kill the study, and they did.

At the hearing, Dr. Arthur Upton, then head of the National Cancer Institute, personally asked me to resubmit the grant proposal to NCI. However, I told him I was reluctant to do so. Unless there were changes in the peer review procedures, another review would merely be a rerun of the first. Although these changes were not made, I decided to resubmit. My rationale was that the peer review process deserved another chance, a fair test of whether it could be made to operate in the public interest.

Some effort at improvement was made, and there were good people on the second panel. However, my point was that it only took one or two site visitors who were determined to kill a program to do the job--it did not require rigging the entire panel.

The second time around there were two members of the radiation protection community there to kill the program. One was from the Oak Ridge National Laboratory, an appendage of the Department of Energy that had been involved in the cutoff of Dr. Mancuso's funds and in the 1978 hearing. The other was from an organization which exemplifies the rule "Don't believe names." This is the Institute of Environmental Medicine. The position of the Institute of Environmental Medicine can be judged from the enthusiastic support that it has received in the Mobil Oil advocacy advertisements. Genuine environmentalist groups are the target of constant attack in these ads, but there is only praise for the Institute of Environmental Medicine. So just as I had predicted, the two site visitors who came to kill the studies did their job more politely, but just as surely.

The application of the third disincentive has cut off all of our funding for the line of research described in this book. Perhaps we had carried this work as far as was necessary to answer the basic questions about low-level radiation hazards, such as the doubling dose for leukemia. There was more to be learned, however. There was also the sadness of saying goodbye to my hard-working and dedicated staff. They were being punished for my crime of leading the program into an area where the federal agencies didn't want us to go. Of course, we were not the only ones who have been punished for this crime.

There are incentives as well as disincentives. Dr. Upton fared rather well. He has since resigned from the NCI and has become the head of the Institute for Environmental Medicine with a raise in salary.

In political action, however, it doesn't pay to dwell on the past. The advice on this is easier for me to give than to follow:

Learn from the past, but look to the future.

What can be learned from our experience with the third disincentive? Just this: There is little prospect for federal support of effective public health research or effective programs for primary prevention of cancer and other diseases as long as the peer review process is used. The system is basically a buddy system. It is extravagantly admired by professionals because they can use it to their own advantage. It destroys accountability to the public. Accountability is something that the health industry does not want and has long managed to avoid.

There is an inherent conflict of interest in the peer review system here. An ever-increasing proportion (probably now a majority) of cases of cancer and other chronic diseases are technogenic, and many of these are iatrogenic. Almost all of the persons in the peer review process, particularly in the upper echelons, have direct or indirect connections with chemical or other industries or with the health industry. They may pay lip service to "prevention;" however, they will not support with dollars any research on primary prevention which might result in restrictions being placed on the chemical or the health industry in order to control health hazards.

Idealistic persons at the site-visit level are brought in for window dressing. The peer review process is actually controlled by "study groups" and other high-level advisory panels. So there is little chance that the public will get innovative or even competent research (or action that will actually save lives) as long as the present peer review system passes out the taxpayer's money.

Changing or reforming ponderous bureaucratic structures such as the National Institutes of Health or the Department of Energy seems just about impossible. The best bet for the public would be to eliminate the present federal agencies and to start all over again.

REFERENCES FROM "THE LIBASSI REPORT"
CITED IN DISCUSSION:
CONFIRMATION OF THE DOUBLING DOSE*

1. The Effects on Populations of Exposure to Low Levels of Ionizing Radiation. Report of the Advisory Committee on the Biological Effects of Ionizing Radiation (BEIR Report), National Academy of Sciences, November 1972.

2. Interagency Task Force on the Health Effects of Ionizing Radiation: Report of the Work Group on Science, June 1979.

3. I. D. J. Bross, Testimony before the Senate Government Affairs Subcommittee on Energy, Nuclear Proliferation, and Federal Services, March 6, 1979.

4. W. H. Foege, Testimony before the Subcommittee on Health and the Environment, U.S. House of Representatives, January 25, 1978.

5. J. L. Lyon, M. R. Klauber, J. W. Gardner, and K. S. Udall, Childhood leukemias associated with fallout from nuclear testing. *New England Journal of Medicine, 300:*397-402 (1979).

6. J. Rotblat, The puzzle of absent effects. *New Scientist, 75*:475-476 (1977).

7. R. A. Conard, Summary of thyroid findings in Marshallese 22 years after exposure to radioactive fallout. In *Radiation-associated Thyroid Carcinoma* (J. deGroot, ed.), Grune and Stratton, New York 1977, pp. 241-257.

8. M. L. Rallison, B. M. Dobyns, F. R. Keating, J. E. Rall, and F. H. Tyler, Thyroid disease in children. A survey of subjects potentially exposed to fallout radiation. *American Journal of Medicine, 56*:457-463 (1974).

9. I. D. J. Bross, Radiation Standards and Public Health. Proceedings of the Second Congressional Seminar on Low-Level Ionizing Radiation, Washington, D.C., February 10, 1978.

10. N. Solomon, U.S. Marines--Nuked in Nagasaki. *The Progessive,* July 1979, pp. 21-27.

11. T. F. Mancuso, A. Stewart, and G. Kneale, Radiation exposures of Hanford workers dying from cancer and other causes. *Health Physics, 33*:369-385 (1977).

12. G. W. Kneale, A. M. Stewart, and T. F. Mancuso, Reanalysis of data relating to the Hanford study of the cancer risks of radiation workers. Proceedings of the Symposium on the Late Biological Effects of Ionizing Radiation held by the International Atomic Energy Agency, Vienna, March 13-17, 1978.

---

*See page 157

13. T. W. Anderson, Radiation exposure of Hanford workers: A critique of the Mancuso, Stewart and Kneale report. *Health Physics, 35*:743-750 (1978).

14. S. Milham, Jr., Occupational mortality in Washington State, 1950-1971. HEW Publication no. (NIOSH)76-175, vol. 1, 1976, pp. 29-30.

15. T. Najarian and T. Colton, Mortality from leukaemia and cancer in shipyard nuclear workers. *Lancet, 1*:1018-1020 (1978).

16. H. J. Evans, K. E. Buckton, G. E. Hamilton, and A. Carothers, Radiation-induced chromosome aberrations in nuclear-dockyard workers. *Nature, 277*:531-534 (1979).

17. V. E. Archer, J. D. Gilliam, and J. K. Wagoner, Respiratory disease mortality among uranium miners. *Annals of the New York Academy of Sciences, 271*:280-293 (1976).

18. N. Kochupillai, I. C. Verma, M. S. Grewal, and V. Ramalingaswami, Down's syndrome and other related abnormalities in an area of high background radiation in coastal Kerala. *Nature, 262*:60-61 (1976).

19. K. J. Johnson, Evaluation of the hazard to residents of areas contaminated with plutonium. Proceedings of the Fourth International Congress of the International Radiation Protection Association, 1977, pp. 243-246.

20. I. D. J. Bross, Adversary science in Aliquippa. *Health Physics, 26*:581-583 (1974).

21. I. D. J. Bross and N. Natarajan, Leukemia from low-level radiation. Identification of susceptible children. *New England Journal of Medicine, 287*:107-110 (1972).

22. I. D. J. Bross and N. Natarajan, Genetic damage from diagnostic radiation. *JAMA, 237*:2399-2401 (1977).

23. I. D. J. Bross, M. Ball, T. Rzepka, and R. E. Laws, Preliminary report on radiation and heart disease. *Journal of Medicine, 9*:3-15 (1978).

24. I. D. J. Bross, Presentation at a public meeting on the low-level effects of ionizing radiation. U.S. Nuclear Regulatory Commission, April 7, 1978.

25. I. D. J. Bross, M. Ball, and S. Falen, A dosage response curve for the one rad range: Adult risks from diagnostic radiation. *American Journal of Public Health, 69*:130-136 (1979).

26. A. Stewart, W. Pennybacker, and R. Barber, Adult leukaemias and diagnostic x-ray. *British Medical Journal, 2*:882-890 (1962).

27. F. W. Gunz and H. R. Atkinson, Medical radiations and leukaemia: A retrospective survey. *British Medical Journal, 1*:389-393 (1964).

28. R. Gibson, S. Graham, A. Lilienfeld, L. Schuman, J. E. Dowd, and M. L. Levin, Irradiation in the epidemiology of leukemia among adults. *Journal of the National Cancer Institute, 48*:301-311 (1972).

29. J. F. Bithell and A. M. Stewart, Prenatal irradiation and childhood malignancy: A review of British data from the Oxford survey. *British Journal of Cancer, 31*:271-287 (1975).

30. B. MacMahon, Prenatal x-ray exposure and childhood cancer. *Journal of the National Cancer Institute, 28*:1173-1191 (1962).

31. S. Graham, M. L. Levin, A. M. Lilienfeld, L. M. Schuman, R. W. Gibson, J. E. Dowd, and L. Hemplemann, Preconception, intrauterine, and post-natal irradiation as related to leukemia. *National Cancer Institute Monographs, 19*:347-371 (1966).

32. G. M. Matanoski, R. Seltser, P. E. Sartwell, E. L. Diamond, and E. Elliott, The current mortality rates of radiologists and other physician specialists: Deaths from all causes and from cancer. *American Journal of Epidemiology, 101*:188-198 (1975).

33. S. H. Macht and P.S. Lawrence, National survey of congenital malformations resulting from exposure to roentgen radiation. *American Journal of Roentgenology, 73*:442-466 (1955).

34. E. L. Diamond, H. Schmerler, and A. M. Lilienfeld, The relationship of intra-uterine radiation to subsequent mortality and development of leukemia in children. *American Journal of Epidemiology, 97*:283-313 (1973).

chapter 12

HOW TO TELL WASHINGTON THE TRUTH

Understanding the role of technical languages and everyday English in public affairs.

PRISONERS OF JARGON

The subtle but dominant role that our languages play in determining what we say and how we act, particularly the role that the technical jargons play in controlling the behavior of their speakers, has already been noted in passing several times. The purpose of this chapter is to go into the linguistic issues in somewhat more detail, but without going into technical aspects. The main purpose is to suggest a few guidelines that the reader might keep in mind. This is a fascinating topic, but anyone who wants to go into it more seriously can find the book-length treatment the topic deserves in my book on scientific strategies for telling the truth.

In an attempt to boil this complex subject down to a simple set of practical guidelines, a list which is provocatively called the "Ten Commandments" was drawn up. It is not a perfect list and certainly not a complete list, but it does cover some points which have proved useful on a number of different occasions. The material in this chapter was presented at a session on "Statistical Problems in Nuclear Regulation" at the 1978 Annual Meeting of the American Statistical Association in San Diego.

---

*This chapter is being reprinted with permission from the *American Statistician,* vol. 34, no. 1, 1980, pp. 34-38.

The "Ten Commandments" were presented as pointers for statisticians who might be testifying as expert witnesses to Congress or to other governmental agencies. However, they apply equally well from the opposite standpoint--from the standpoint of nonexperts who are trying to judge whether so-called "experts" are telling the truth. If there is one thing that this list has in common with the original Ten Commandments, it is that these guidelines are often ignored or violated. Moreover, the very persons who are most prone to violate them are the elitists in science or in official positions who consider themselves intellectually and ethically superior to ordinary mortals. Those who are most inclined to flaunt their own "superiority" are also most inclined to give false assurances of safety and otherwise to lie to the public and its representatives.

A number of representatives of regulatory agencies, or of in-house or outside laboratories supported by funds from the Department of Energy or the Nuclear Regulatory Commission, were present at this talk and, suprisingly perhaps, they were the most positively reactive segment of the audience to these themes. A point which the public can use to its advantage is this: In these public health issues there are "good guys" and "bad guys," but none of the "bad guys" likes to have this public image.

TESTIMONY BY SCIENTISTS

At some point in his or her career, a scientist may have the task of presenting findings to some governmental body, possibly a congressional committee, regulatory agency, or perhaps as part of a judicial process. All of us may want to tell the truth, but it isn't always easy to do this under the conditions that exist in some governmental agencies. For instance, a regulatory agency may not want to hear the truth if it happens to go against the official doctrines or policy of that agency. There are certain gimmicks that may be used to restrict the individual's opportunity to tell the truth--or at least to offset, muddle, or confuse whatever the person has to say. Sometimes the statistician may be part of a group or committee. He or she may then be subject to heavy pressures of uniformity or conformity from other persons who seem to know what they are doing. Statisticians are not the most assertive persons, and they tend to go along with what seems to be the majority. This may, in fact, not be the will of the majority, but rather the will of one or two persons who are planted on the committee for the specific purpose of influencing its decision.

Not too many people have had a great deal of first-hand experience in this area and have testified to Congress, courts, or regulatory agencies. In recent years, I have had a lot more experience than most other persons, a lot more experience than I really wanted. What I would like to do today is to pass on to you some of the lessons I have learned from these experiences and also from my past studies on the use of natural languages in human affairs.

Not many scientists or statisticians really know very much about the nature of natural languages. Most of the time we tend to stay inside the technical jargon of our specialties. Many of us are more at home with the precision of mathematical languages, either algebraic or arithmetic, than we are with the apparent imprecision that we encounter with plain English. Nevertheless, in dealing with the public and its representatives, the most effective language for communicating information is one which is close to ordinary English.

On several past occasions, I have given a list of what I call the "Ten Commandments" for speaking to the public about statistics. I will again use this list in this presentation. However, the examples I will give of the use of the commandments will be entirely different from my previous ones. This time they will involve my past experience in trying to communicate scientific and statistical ideas to the federal government.

1. DON'T SPEAK A FOREIGN LANGUAGE; SPEAK PLAIN ENGLISH

I have already brought this point up because it is one of the hardest of the commandments to follow. For a variety of reasons, some of us are afraid, reluctant, unwilling, or unable to translate our technical jargons into plain English. When we speak in technical jargon we sound like an expert. We lose this advantage when we speak everyday language. Our expertise then must come out in what we say instead of how we say it. Another difficulty is that, when a statement is translated from a technical jargon to ordinary English, something may be lost in the translation. Some of the precision and accuracy which is possible in a technical jargon will be lost when we use the less exact terminology of ordinary English.

This may expose one to criticism from those who make a fetish of technical definitions. They will complain that this is not a very professional way to speak. What we have to realize, however, is that there is a trade-off in this translation. We must be prepared to sacrifice some precision of statement so that what we do say makes more sense to the public. It really does not do a lot of good to make a very precise statement when no one understands what the statement says.

Here I would make a distinction between being a professional and being a scientist. There are certain professional customs common to most professions--medicine, law, or science. These professional characteristics are in no sense scientific characteristics. Rather, they represent sociological attitudes of in-groups, particularly in-groups that speak a technical language. It may be unprofessional to speak plain English, but it is not unscientific to do so.

If you do try to speak plain English to legislators or judicial commissions, then there are going to be a number of difficulties, pitfalls, and mistakes which you may fall into. The next commandment can help you to avoid one common mistake.

2. DON'T TALK SHOP, TALK SENSE

A recent example occurred during hearings of the Subcommittee of the House on Health and Environment which was chaired by Congressman Paul Rogers of Florida. Dr. Tim Lee Carter, a representative from Kentucky, was concerned with the effects on soldiers exposed to fallout at the Big Smoky weapons test 20 years ago. He asked each and every statistician and epidemiologist at the hearing the following question:

Given that 447 soldiers who were exposed at Smoky have been traced, and given that eight of them have developed or died of leukemia, is this a "statistically significant" result? The Congressman offered the further information that about four per hundred thousand per year was the leukemia rate in the 40-year age group.

The Congressman got almost a dozen different answers from the witnesses. Nearly all of them started out with a long and complicated set of qualifications, conditions, or hedges on the subsequent statement. According to Dr. Carter, all of the statisticians eventually ended up saying "yes." However, most of them were reluctant or incapable of simply saying yes. One witness, for instance, said "I haven't got my alpha." However, Congress was not impressed with the witnesses who did everything possible to avoid giving a straightforward answer to a straightforward question, by going into all sorts of technicalities.

3. AVOID WORDS THAT MEAN DIFFERENT THINGS IN DIFFERENT JARGONS

Think what the word will mean to the audience. The regulatory agencies in general, and the Nuclear Regulatory Commission in particular, have been guilty of using terms in special technical senses which are very misleading to the public and to the representatives of the public. Sometimes people are surprised to find out what the "technical" definitions of terms actually are. For instance, the NRC sets 5 rems per year (lifetime average) as the "permissible" exposure for nuclear workers. A rem is a measure of the amount of radiation similar here to another measure, the rad. It is natural to assume that this dose is, in some sense, a "safe" dose. Publications of the Atomic Energy Commission (AEC) and the Nuclear Regulatory Commission strongly suggest this.

So when I presented evidence to the NRC that the doubling dose for leukemia in humans is probably less than 5 rems, and therefore that the level is clearly not safe, one might suppose that the NRC would hasten to lower its "permissible level" to, say, 0.5 rems.

However, it was argued that the new findings were not relevant because the "permissible level" had nothing to do with safety!

while the public thinks that the government would not permit any level of exposure which is not safe, technologists can argue that "permissible" is defined in terms of engineering feasibility. To use definitions that contradict ordinary usage is, in fact, not much different from lying to the public.

4. DON'T FOCUS ON ISOLATED WORDS; CONCENTRATE ON THE MESSAGE

This commandment may seem to contradict the previous one, something not uncommon with general guidelines. The proverb "Look before you leap" contradicts "He who hesitates is lost." When we use any guideline we must recognize that there are different circumstances. The guideline chosen has to be appropriate for the circumstances. Sometimes it is necessary to pay attention to isolated words, because words such as "permissible" may be the crux of the issue. Other times, the object is to get a general message across. Too much focus on the specific words can interfere with this effort.

For instance, one of the important new findings that I had to communicate at the NRC hearing was that the "doubling dose" for leukemia was likely to be less than the dose of 5 rems permitted each year. There are a lot of technicalities involved in the concept of a "doubling dose." Some scientists would feel it was absolutely necessary to make long and complicated statements in very technical terms about the specific estimates. However, the technicalities are almost unintelligible to most people. The real point is not the particular number, but the fact that the estimate probably lies inside the currently permissible level. Obviously, an exposure that doubles the chance of a malignant disease is not a "safe" exposure. However, it is easy to lose this central point if one gets into the technicalities of dosimetry.

5. DON'T EXPECT TO RELAY A MESSAGE IN ONE SHOT; BE CAREFULLY REDUNDANT

This is one commandment which contradicts what is commonly drilled into graduate students. Everyone who has had a technical paper severely criticized or rejected because of redundancy--the repetition of a statement a few paragraphs or pages after the original statement--is sensitized to this point. However, redundancy is not the terrible thing that it is sometimes made out to be. In natural language there is a lot of redundancy. It serves a useful communication purpose.

When someone hears about a complicated technical issue for the first time, it is often difficult to remember all the information. If the information is repeated at points where it is particularly relevant, the improved communication can more than offset the irritation that redundancy sometimes produces. Mathematicians are likely to react very strongly against redundancy because one of the characteristics of mathematical languages is that they tend to be nonredundant. However, a style which may be fine for communication between mathematicians is not necessarily satisfactory for communicating with the public or its representatives.

In all communication there is the risk that what will work with one person will not work with another, and vice versa. There is always a chance that professionals in an audience may be affronted if the message comes in plain English that does not cater to the professional vanity. With mixed audiences of professionals and the public, it is just about impossible to please everybody.

6. DON'T IMITATE THE RHETORIC OTHERS USE; SPEAK NATURALLY

There is probably no greater mistake that a statistician or any scientist can make in speaking to professionals outside his or her specialty than to use the language of another specialty and to use it incorrectly. This commandment makes this point. To use unfamiliar legal jargon when talking to lawyers, or to mispronounce medical terms when talking to doctors, is the surest and quickest way to lose all credibility with the audience.

Languages are used partly as identification signals so that the person in a specialty is not fooled by an outsider who tries to imitate the language of the specialty. In statistics, almost all of us have had the experience of hearing persons who really know nothing about statistics attempt to sound like expert statisticians. These phonies may fool the public, but they don't fool a professional.

In the communication of material that is highly technical, particularly when other speakers have dwelt on technicalities, it is difficult to avoid imitating the language that seems to be expected. It is particularly true for someone who is new to the business of testifying before governmental panels. Witnesses will often speak such a very technical-sounding language that a newcomer may feel that use of anything but his or her specialized jargon would raise doubts about his or her competence or expertise in the minds of the listeners. I would not deny that this is a risk. On the other hand, experienced panels have heard all sorts of nonsense in technical-sounding double-talk. They are not overly impressed by the sounds of any language. If the panel is competent, it will pay attention to the sense of what is being said and not the way of saying it.

At a hearing where there is a transcript being made, the speaker should be aware that the message is going to be read as well as heard. The speaker has to try to make statements that read well and also sound well. This is not easy because there are distinct differences in the usage of written English and spoken English. The speaker should be aware of these differences. In general, knowing more about the nature of the natural language is something which will be a valuable background for anyone who wants to tell the truth in Washington--or for that matter anywhere else.

Therefore, let me say just a little about the science of language, which is called "modern linguistics." Linguistics has its own jargon, so it isn't easy for an outsider to read this literature. I have given some introduction to modern linguistics and a listing of readable textbooks in my book, *Scientific Strategies in Human Affairs: To Tell the Truth* (Exposition Press, 1975).

Please do not confuse the science of modern linguistics with what you have been taught in school about the English language or other natural languages. Most of what has been said and written about English is nothing more than prescientific nonsense. For example, educators often proclaim how chaotic and poorly organized the English language is. Bunk! All natural languages are well-organized structures or they could not serve the function of communication. What seems to be chaos is simply a misunderstanding of how languages actually work.

7. DON'T SPOUT TEXTBOOK DOGMAS

Perhaps one reason why statisticians may underestimate the importance of natural language in the difficult business of telling the truth--of communicating an accurate and precise message about the state of affairs in the world about us--is that we have a tendency to concentrate on one part of the communication process. We tend to focus on the beginning of the process, on finding the truth. This is a task where our statistical expertise can come in effectively. However, simply finding the truth is not enough. For the public to benefit from the truth, it has to be communicated. It is in this communication process that the role of language is especially important. All of us, and I don't exclude myself, need to know more about how to effectively communicate the truth to the public.

Often, we are overly defensive in our professional communications. For example, we make "safe" statements from textbooks or give elaborate definitions of concepts, which add very little to the discussion but give the impression of expertise and allow us to say things which may be irrelevant but are easily defensible. On the other hand, when we venture into our own work, our own experience, we are naturally afraid that someone will catch us or criticize us or make us look bad. This is not an idle fear because in public issues it is a standard tactic to try to discredit those who speak on behalf of the public on, say, environmental hazards. In spite of this, I strongly recommend that you tell what you have learned for yourself and not what is the prevailing textbook dogma.

8. DON'T BE A HIGH-PRESSURE SALESMAN

It may be more dangerous to talk about your own work than to assert the platitudes and tautologies which are popular in statistics as in other areas. But if you are going to speak the truth you must be prepared to stick your neck out, and possibly have your opponents try to chop it off. This is part of your responsibility to the public as a professional. Unpleasant as it may be at times, there is no legitimate way of avoiding these responsibilities.

There is often a strong temptation in testifying in Washington and elsewhere to put a good word in for something near and dear to one's heart--a plug for one's subspecialty, for a particular technology or methodology, or for an area of research or class of institutions which just happens to be one's own. While professional folkways don't frown on this kind of self-advertising, it tends to come across as a self-serving statement which can discredit other and more objective testimony. Congressional committees, federal agencies, and other groups hear mostly from lobbyists, and discount what they hear from these sources. It doesn't help to sound like a lobbyist. So my advice is: Don't be a high-pressure salesman; the public isn't buying these days.

9. SILENCE CAN BE GOLDEN

The next commandment is one of the most important for telling the truth: If you don't know what you are talking about, don't talk. All of us have our fair share of ego (some a good bit more than this). It isn't easy to say "I don't know" when one is testifying as an expert. However, it is not really a crime for anyone to admit that there are areas where he or she does not know the answers. An honest admission can sometimes (but not always) add credibility to the rest of the testimony.

The real problem is that we have learned some bad habits--a variety of linguistic gimmicks which are built into our professional jargons. These enable us to give the impression that we know what we are talking about, when in fact we don't. Any lecturer in statistics or other subjects in a university has acquired a bag of tricks for dealing with difficult or embarrassing questions from students--for putting the "uninformed public" in its place. There is a temptation (not often resisted) to use these gimmicks in testifying to regulatory or other public agencies when awkward questions come up. In professional jargons, it is relatively safe and easy to lie to the public by using these tricks in technical languages. However, when professionals do this, they are betraying the public trust that has been placed in them.

There is one place where it is especially easy for persons who do not know what they are talking about to sound as if they do. This is when criticizing the work of other statisticians or scientists using statistics. There are a very large number of standard put-downs which are in wide use in academia. They are criticisms by innuendo rather than statements that show specific errors in the data, methods, or findings.

What can be done to raise the depressingly low standards of "statistical criticism" and "scientific controversy" that now exist? What can be done to stop this kind of nonsense from delaying or blocking essential public health action? The cigarettes-and-lung-cancer issue is a good example of how statisticians and others, who played ego-serving language games in this controversy, have caused hundreds of thousands of unnecessary deaths and disabilities over the past 25 years. Is there any way to resolve these statistical issues more quickly and scientifically? One device which, I believe, has promise is called "Adversary Science."

This adapts the adversary procedure used in courtroom trials to scientific questions. There is a spokesman (a scientist rather than a lawyer) to present the case for each side. There is a neutral chairman presiding. The arguments are made, as to a jury, in plain English. The presentations, rebuttals, cross-examinations, and summations are recorded as transcripts. This allows the public and its representatives to review the testimony and make the final decisions like a jury. The NRC hearing on April 7, 1978 was the first Washington test of this new format.

While not ideal, this Adversary Science format did allow the controversial question of the hazards of low levels of ionizing radiation to get a relatively clear public answer. It gave both participants the opportunity to put the last commandment into effect: Tell the public the truth, the whole truth, and nothing but the truth to the best of your ability.

10. NOTHING BUT THE TRUTH

This is, of course, the most important commandment. The other nine really are guidelines for implementing this difficult tenth commandment.

A natural question is: Can the public be expected to understand the technical, statistical, or scientific issues that may arise? I believe there is a better opportunity for this in an Adversary Science format than in traditional formats. An actual example may show why.

Part of my NRC testimony was the presentation of the first dosage response curves ever constructed for the 1-rad exposure range--the range between 100 millirads and 10 rads. All previous curves have used linear extrapolation from data on much higher exposure (e.g., hundreds of rads). This was the first curve ever constructed directly from data on men exposed to ordinary diagnostic radiation in the 1-rad range.

Without going into the underlying mathematical model, let me simply note that one parameter of the model was "the proportion of the population affected" by a given dosage. In the prepared critique, it was objected that this was not a proper y-axis for dosage response curves, but elsewhere it was indicated that the "percent excess leukemia risk" would be appropriate. This is the kind of technical issue which the public might not ordinarily be expected to resolve. If the experts disagree, the public often feels it has no hope of resolving the issue.

The issue was not actually a difficult one because for all practical purposes the response variable used in the presentation was simply proportional to the one tacitly accepted by the critic. Since the NRC asked me for a written clarification of this specific point, I could make the point clear in four languages--algebraically, arithmetically, graphically, and verbally. While the critic and the "radiation protection community" could refuse to listen to the argument in any language, "concerned citizens" could read the written transcript and decide for themselves. I think that the public can see this point: If a given variable is an appropriate measure for response, dividing it by 9 does not make it inappropriate.

In sum, I believe that a statistician or a scientist can tell the truth in Washington and to the public, but that it takes a strong commitment to do so. Much hard work and preparation is needed to produce statements which are accurate and comprehensible. The speaker should have a sense of responsibility to the public. In the past 30 years, every economically important public health issue has been tied up by pointless and unnecessary "controversies" that blocked effective action to protect the public. The statisticians, scientists, and physicians who created or prolonged these "controversies" cannot escape some responsibility for the deaths of these human beings. Playing language games with words like "cause" may be a favorite academic game, but it is irresponsible to do this when someone's health and safety are at stake.

To avoid future repetitions of past folly, professionals should watch their language much more closely in the future. Unless professionals start telling the truth to the public in a comprehensible way, the public is going to stop listening to what we have to say.

THE TEN COMMANDMENTS

In summary, the ten commandments for speaking to the public about statistics and science are:

1. Don't speak a foreign language; speak plain English.
2. Don't talk shop, talk sense.
3. Avoid words which mean different things in different jargons. Think of what the word will mean to the audience.
4. Don't focus on isolated words; concentrate on the message.
5. Don't expect to send a message in one shot; be carefully redundant.

6. Don't imitate the rhetoric others use; speak naturally.
7. Don't spout textbook dogmas; speak from first-hand experience.
8. Don't be a high-pressure salesman; the public isn't buying these days.
9. If you don't know what you are talking about, don't talk.
10. Tell the public the truth, the whole truth, and nothing but the truth to the best of your ability.

chapter 13

WEIGHING BENEFITS AGAINST COSTS

Does risk-benefit or cost-benefit analysis help or hurt in resolving public health controversies?

SHOULD YOU BE REASONABLE?

It seems only reasonable that when decisions about the large-scale deployment of new technologies are to be made, they should be based on a careful weighing of the benefits against the costs. There are different ways of doing the calculations, and various names are used such as "cost-benefit analysis," "policy analysis," and "statistical decision-making." In theory, the idea is the same--to get an estimate of the potential benefits (and chances of realizing these benefits) and the potential costs, and then to compare the two to make the decisions on deployment. In theory, it is a good idea. Unfortunately, in practice it has generally worked out badly for the public.

In most public health issues where the debate has moved into the area of benefits and costs, I have noticed that the benefits go to the promoters of technologies, while the public pays the costs in death or disability. For example, the debate on the reduction of the permissible level from 5 rads (rems) did not end with the April 7, 1978 hearing. The impact, if any, of the factual evidence was to divert the debate away from the health issue into the benefit-cost area. This is the usual pattern when the technologists realize that they can no longer fool the public by categorically denying the hazard. The next step is to confuse the public with the cost-benefit issues.

Confusion comes easily because there is often much less factual evidence for assessing costs and benefits overall than there is for assessing the health hazards. So what is likely to happen--and this sometimes seems to be the only point of cost-benefit analysis by technologists--is that the general fogginess obscures the hazards. In the continuing debate on cutting the 5-rad level, the argument has switched from the hazards at this level to the added cost to the nuclear industry of operating at the 0.5-rad level.

When the debate veers in this direction, the public is asked to "be reasonable." But to high-pressure salesmen, "be reasonable" is a euphemism for "be gullible." So in being reasonable the first question you should ask is: "Who gets the benefits? Who pays the costs?" Unless the persons who may be harmed are the ones to get the benefits, the whole thing is, ethically, nothing but a fraud.

COST-BENEFITS OF REGULATION

It is easy to lose one's way in the ethical morass one gets into in weighing human lives in the balance, and it really is more reasonable to say: I believe that "Thou shalt not kill" applies to technologists. Technology should not be deployed in ways which will result in human beings being killed or seriously harmed. If regulatory agencies, such as the NRC, had this viewpoint, the evidence I presented would have ended the debate.

Even someone like Dr. Karl Z. Morgan, whom I respect, gets into this morass in cost-benefit considerations. He argued against the reduction to the 0.5-rad level on the grounds that the nuclear industry, to operate economically at this level, would resort to "burn out." This means hiring large numbers of persons, using them in "hot" areas until they got up to the permissible level, and then firing them. This burn out could be worse from the overall public health standpoint than higher levels of exposure to a smaller number of persons.

The point about burn out is a good one. Indeed, in the NRC question period, I took the time to point out to the NRC that this dangerous practice should have been banned long ago. However, it is not really a good argument against a protective regulation to say that it might be evaded in hazardous ways.

There is always a possibility of malevolent counteraction to any action to protect the public. This simply means that regulation requires enforcement. But one gets into this fantasy world of "if" and "counter-if" when one gets away from the simple facts about health hazards.

Indeed, as a critic of Dr. Morgan pointed out, there is this paradox: If, for whatever reason, at the end one is going to disregard the actual health hazard as a basis for setting the permissible level, why bring it up at all?

MICKEY MOUSE ARITHMETIC

When technologists do calculations about costs or benefits using numbers that have little if any relationship to the actual costs or benefits, this is what I call "Mickey Mouse arithmetic." Looking at a Mickey Mouse cartoon is no way to learn the biological facts about mice. Looking at Mickey Mouse arithmetic is no way to learn about the action situation in a health issue.

The calculations of benefit-to-cost balances inherently involve human values--including the kind of values which come under "ethics" as well as those which can be measured in dollars. The value system of the person doing the calculations (or the persons for whom the calculations are being done) comes into the arithmetic whether this is done consciously or unconsciously.

The point is that any real action situation is so complex that there are often no good numerical estimates for most of the values, probabilities, or other parameters that will come into the problem. This means that the numbers that are used are going to be chosen in a highly subjective way. These choices will reflect the personal prejudices of the calculator. For economists whose stock-in-trade is this kind of Mickey Mouse calculation, this does not pose any problem. For an honest scientist, however, this kind of guesswork is likely to be rejected as nonsense. Sometimes the whole idea of trying to strike a balance in complex situations will be rejected as well.

The counterargument to this complete rejection is: Like it or not, we live in this world and we have to make decisions in it. In the last section in this chapter we will take a look at the possibilities for a genuine scientific approach to decision-making, at the prospects for a new technology for the safe and effective deployment of technology ("metatechnology"). At this point, we can simply note that the usual follow-up to the counterargument is fallacious. This line goes: You should accept the [Mickey Mouse] cost-benefit calculations because they are a more rational way to make decisions than no calculations at all.

This just isn't true. The Mickey Mouse calculations are a misuse of arithmetic in the same way that the rationalization of an evil is a misuse of reason. Most policy analysis, systems analysis, or cost-benefit analysis is not for the purpose of making a decision. The calculations are for the purpose of "rationalizing" a decision that has already been made. It is just as easy to lie in mathematical languages as in any other languages. Mickey Mouse calculations are simply a means of lying in a language which seems "scientific" and which the public doesn't understand.

The only effective strategy for dealing with Mickey Mouse arithmetic is to identify it as such and then throw it out completely. It is a serious strategic mistake to try to "correct" the arithmetic. This is often a temptation--to challenge the numbers and to change them so as to make the balance come out the opposite way. This plays into the hands of those whose goal is to start a "controversy" to block public health action. This is a strategic mistake that interveners often make in "environmental impact" calculations. When these are nonsense to begin with, changing the numbers cannot change the fact that the whole basis for the calculations was unrealistic and unscientific.

Concerned citizens should not only shun this nonsense, they should shun the technologists *on both sides* who engage in this Mickey Mouse arithmetic. With the Rasmussen report, for instance, the entire basis for the calculations was nonsensical. The counter-calculations simply misled the public into thinking there was something to the arithmetic. This can provide an ideal way for technologists to pose as great friends of the public while acting in collusion with their colleagues. This is a no-lose proposition for the grand-standing scientists, but a no-win situation for the public.

If these judgments on the cost-benefit game seem unduly harsh, let us consider in some detail an actual case history. This will allow you to see for yourself how this game is played, why it is played, and who are the prestigious players that may take a hand in this game.

THE SNOWBALL EFFECT

This is the story of how a young radiologist made a mistake in his calculations--a mistake which was so big and so obvious that it would stand out like an unexpected extra $10,000 in your bankbook--and how this mistake was propagated by organizations of the highest reputation in medicine, science, business, and journalism. It wasn't just an abstract mistake; it was an error which posed a threat to the health and safety of 270,000 American women. So it is instructive to follow what happened here step by step as an unusually clear-cut example of how a "Snowball Effect" can convert a completely false claim for the safety of a technology into a widely accepted "scientific fact." The Snowball Effect is produced by echoing and re-echoing the false statement back and forth between prestigious scientific, technical, or medical organizations and the mass media. While the Snowball Effect is not the only gimmick for keeping the truth about technology from the public, an understanding of how this gimmick works can give one some protection against this and other gimmicks.

The young radiologist was calculating what is called a "benefit-to-cost ratio" for the women in the huge American Cancer Society program for the mass screening by mammography. Nowadays it is fashionable to calculate this ratio because, when the benefits far outweigh the costs, this can be used as an argument to get money from Washington. In this instance there is at least $6 million a year riding on this benefit-to-cost ratio. However, the ratio the young radiologist was calculating was a little different from the usual ratio because both the benefits and costs were being measured in terms of the number of deaths from breast cancer rather than dollars. This makes the calculations a little tricky. This is not something that radiologists are accustomed to doing. So it isn't surprising that this radiologist made a mistake.

In the later cover-up of the mistake, all parties concerned tried the usual ploy that a "scientific controversy" was involved. In fact, the mistake was as clear-cut and simply arithmetic as a mistake in a checkbook. So let us begin with the mistake itself. The young radiologist was redoing or revising the previous calculation of the benefit-to-cost ratio for this program that had been made by Dr. John Bailar, editor of the *Journal of the National Cancer Institute* and an epidemiologist and biostatistician with extensive experience in this kind of calculation. Bailar used in his calculations a number which he took from the report of the National Academy of Science on the Biological Effects of Ionizing Radiation. The number is simple, but the units are a little complicated: six breast cancers per million rads per year. To calculate the number of breast cancers which would be *induced* by the x-ray delivered in the mammographic examination (which is the "cost" here) takes only simple arithmetic. What you do is multiply the six by the number of women in the program, by the average amount of radiation per examination (about 2 rads), and by the number of years at risk.

The radiologist simply forgot to multiply by the last factor, the number of years at risk. Without trying to go into details of how this number of years is obtained, it is sufficient to note that it is calculated using the life expectation. Life expectation depends on age, but for the women in this study it varies from 20 to 40. Taking 20 as a round number for the order of magnitude of the number of years for present purposes, it can be seen that failure to make this multiplication makes a whopping error in the estimate of the costs. There were also errors in the estimate of the benefits so that the estimate that the radiologist obtained was off by a factor of about 100--a glaring error.

This is not a question of a "scientific controversy;" this is a question of a gross arithmetic error that any competent expert in this area should spot immediately. So we have here an accidental but scientific test of the competence of the experts in the various organizations that eventually got involved in this episode.

PROPAGATION OF ERROR

The young radiologist came out with a benefit-to-cost ratio of 125, a distinctive number. Like a fingerprint, it provides positive identification in all the various press releases, science news stories, and advertisements that were to stem from this mistake. Naturally, a figure of 125-to-1 looks a lot better to a radiologist than the figure of 1-to-1 that Bailar originally reported to the American Association for Cancer Research in May 1975. When the arithmetic is corrected, however, the two estimates agree rather nicely at about 1-to-1.

A 1-to-1 benefit-to-cost ratio is a discouraging one. Translated from jargon to plain English, it says that this costly mass screening program is going to produce as many deaths from breast cancer as it is likely to avert. When a Jack Anderson column reported Bailar's findings, there was an immediate flap at the American Cancer Society which was running the program and at the National Cancer Institute which was paying for it. What produced the panic was the thought: What will happen to the program if the women find out that the x-ray is doing them as much harm as good?

The 125-to-1 figure came as a godsend to the harried administrators at the American Cancer Society, and they were not about to look a gift horse in the mouth. Here was a total refutation of Bailar's careful work, and it served its purpose well. Bailar's message was drowned out by the noise, and the program rolled on. So it is worth looking a little more closely at how an outright mistake was transformed into an unquestioned "scientific finding."

The American College of Radiology and the American Cancer Society set the stage for the debut of the 125-to-1 figure. The theme of the well-orchestrated conference was reflected in the headline for the *New York Times* story of November 7, 1975 by Jane Brody: "Breast Cancer Project Called Success." The press releases amounted to an advertisement for the supposed benefits and safety of a mass use of an x-ray technology by the two organizations with the most direct stake in the success of this technology. The *New York Times* and other newspapers may not publish free advertising in other areas, but in science and technology and medicine, a large part of what passes for news releases consists of promotional material.

That the American Cancer Society would bet on a gift horse coming in at 125-to-1 odds in an effort to save a faltering multi-million-dollar-per-year program is hardly surprising. But what brought the prestigious American College of Radiology into the act? The answer: Bailar's findings were even more embarrassing to this medical organization than to the American Cancer Society. Under heavy pressure from the American Cancer Society, a stacked committee of radiologists had given a blanket endorsement to the mass screening by mammography. British and Canadian radiologists had opposed the mass screening of women under 50. Most of the American radiologists I've talked to have told me privately that they don't like mass screening under 50. The leadership rammed through the endorsement, and Bailar's work put them in an uncomfortable position. So once again, no one was looking a gift horse in the mouth.

BRINGING IN THE ELITE

Gross errors are an old story with benefit-to-cost ratios. The benefits of technology are consistently overestimated and the costs consistently underestimated. It would seem that while errors may be made originally in both directions, the unfavorable errors are searched out and corrected while the favorable errors are cherished and preserved.

What happened next illustrates the Snowball Effect in the propagation of an error. When press releases appeared in the *New York Times* and other newspapers as science news, the 125-to-1 figure carried the authority of a newspaper with a proud reputation for accuracy. The figure is off by a factor of 100, but who could imagine that the *New York Times* would let such a mistake get by? But then what science writer would suspect the American Cancer Society and the American College of Radiology of issuing such a completely false report on a matter immediately affecting the health and safety of 270,000 American women?

The triple endorsement converted the 125-to-1 figure from a gross error to a certified "fact." This, in turn, opened new channels for the propagation of the error. The Kodak corporation manufactures a new film for mammography that cuts the x-ray exposure drastically--an important technological advance. Impressed by the *New York Times* report, they decided to use the 125-to-1 figure and the other erroneous material in an advertisement on the subject of breast cancer and x-ray. The ad appeared on the back cover of *Science* on July 2, 1976--many months later and after ample time for correction of the original mistake. In effect, the American Association for the Advancement of Science and Kodak are added to the list of organizations certifying this completely false "fact." The ad appeared still later in the *Scientific American*.

The Snowball Effect rolls right over common sense and ordinary care and caution. When I spoke to the executives at Kodak, the reaction was: "How could we have any doubt of the 125-to-1 figure when the story was in the *New York Times*?" When I spoke to the managing editor of *Science*, his reaction was: "Why should we question an ad when it comes from such a reliable company as Kodak?" When I spoke to the *Science* advertising people, they explained that there was no check made on the ads going into *Science* because there was so rarely anything wrong with them. But suppose someone had thought of checking. Could they have had a warning that the 125-to-1 was all wrong?

The answer is that a few hours of library research or a couple of phone calls would have given ample warning. However, despite the reputations for accuracy of the various parties to the Snowball Effect, no one really took the trouble to check out the information. In a way, the story in the *New York Times* on November 7, 1975, carried a warning. The paragraph just before the 125-to-1 started out, "Recently some doctors have criticized the breast cancer screening program. . . ." At the very least, there was a warning that the figure was "controversial" and that further checking was mandatory.

A call to the *New York Times* could have produced the further information that on March 27, 1976, there was another story by Jane Brody on the same subject. This story came out of the show that the American Cancer Society puts on each year for the science writers. In addition to wining and dining the reporters, the American Cancer Society puts on display "noted scientists" who occasionally report on actual research they have done, but more often are promoting some particular program or medical technology that an in-group happens to be pushing. This time they were promoting their own mass screening program and defending it against increasingly heavy criticism. In particular, the screening of women under 50 was now becoming the focus of attention. When Bailar's approach is used separately for the women under 50 and those over 50, then the 1-to-1 benefit-to-cost ratio is further clarified. The screening over 50 becomes more favorable, while the screening under 50 becomes totally indefensible.

SCIENCE REPORTING

The American Cancer Society used its wingding for the science reporters to again try to drown out Bailar's warning message with a lot of static. This time, however, the 125-to-1 figure was deflated by a factor of 20 or 25 (the error of multiplication) The new figure was 5 or 6 to 1. While this is still wrong, anyone who had bothered to check out the 125-to-1 claim would have been warned about the gross error in the calculation. The *New York Times* headline now read: "Benefits, Radiation Risks in Breast Cancer Debated."

In political or commercial reporting, the claims that are made at staged media events, such as this American Cancer Society affair, are viewed by experienced newswriters with some skepticism. Also, there is some awareness that accepting favors of any kind from an organization with an ax to grind involves a reporter in a conflict-of-interest situation. No one is bribed by a few free meals, but the benevolent feeling that may be induced by the largesse is a subtle form of influence.

Unfortunately, many science reporters publish press releases almost verbatim, with little effort to check out the statements, even though the statements are coming from organizations such as the federal science agencies (NASA, DOE, etc.) that have good reasons to slant a release to favor some program or technology that the agency is pushing. The Snowball Effect is largely possible because of the simple faith (or gullibility) of many reporters in this area. The need for tough, investigative reporting is just as great in covering science and technology as in covering business or politics or any other human enterprise. However, it is noteworthy that Bailar's warning appeared in the column of an investigative reporter from the political area. Science reporters shunned this story.

It is to Jane Brody's credit, however, that she did name Bailar in the March 27 story (instead of "some doctors") and, what is even more important, she pointed out the lack of informed consent in this program. Anyone from Kodak, *Science,* or even the American College of Radiology would have had ample warning that the 125-to-1 figure was all wrong and would have a place to go--namely to Bailar--for an explanation of just why the 125-to-1 was wrong.

Anyone can make a mistake. I can sympathize with the young radiologist because when I first calculated a benefit-to-cost ratio for the women under 50 (where the benefits are least and the costs are greatest), I made the same mistake. But a good test of the reliability of these respected sources of information is their willingness to *correct a mistake*.

THE VICTIMS OF THE CRIME

No one likes to publicly admit that he or she goofed. For these prestigious organizations, telling the truth is even more embarrassing since it affects their credibility. Under these special circumstances, however, telling the truth was not just an awkward necessity; it was an inescapable moral obligation.

The Snowball Effect in the propagation of error that has been described here has an innocent and inadvertent aspect--well-meaning persons falling into a trap that was totally unexpected. But unfortunately, there is another, darker aspect to this entire episode. For the women under 50, in round numbers 130,000 women, this mass screening program of the American Cancer Society represents one of the largest human experiments in medical history. The Health Insurance Plan study used to justify the program had found no visible benefit to women under 50. The screening in this group was advocated as a worthwhile experiment because (a) technological improvements in mammography raised the hope of benefits to these women under 50 despite the negative findings earlier, and (b) the mammography had virtually no hazard. It is doubtful whether this was a worthwhile experiment, but there is no question on one point: It was an experiment on human subjects.

Nor is there any question on another point: This experiment by the American Cancer Society was not in compliance with the federal laws or the National Cancer Institute guidelines that are designed to protect the human subjects in medical experiments. The women should have had informed-consent procedures prior to the x-ray which would have informed them of the hazards. If Dr. Bailar's benefit-to-cost ratio had been explained to these women--with the clear implication that the x-ray was doing as much harm as good--it is doubtful that there would have been 270,000 volunteers.

For this program to operate, it was necessary to keep the truth about the hazards of the x-ray from the human subjects of the experiment. Under these circumstances, the publication of the entirely erroneous 125-to-1 figure, however inadvertently, had the effect of keeping the truth from the human subjects of the experiment.

This, then, is not an abstract scientific argument about whether 125-to-1 is a mistake or not. There was a serious violation of the federal laws for the protection of human subjects of medical experiments. The civil rights of at least 130,000 American women had been violated. Particularly in the women under 50, where the benefit-to-cost ratio is especially adverse, there was a serious danger to the health and safety of the human subjects. The prestigious organizations that goofed not only had a responsibility to tell the truth for its own sake, they had a responsibility to tell the truth for the sake of the women.

Note that what has "naturally" occurred here constitutes an interesting scientific experiment on an ethical issue. All parties to the 125-to-1 mistake were confronted with an ethical problem that required some action. Evasion or avoidance of corrective action is itself a course of action here. Accordingly, I personally contacted the organizations involved in the Kodak ad by telephone and for *Science* and the *New York Times* by phone and letter (at their request).

On my initial contact, I received assurances from all that they had only the best of intentions in the publication of the ad. From *Science* and Kodak, I got some assurance that the ad would not be repeated. From the young lady I spoke to in the science news unit of the *New York Times,* there was only the assurance that the *Times* would never be this inaccurate. But no one suggested that there would be any public acknowledgment of error, any retraction, or any effort to set the record straight. None of the parties indicated any intention to communicate to the women in this experiment the truth about this benefit-to-cost ratio, and none of them did so.

THE NEO-JESUITS

One of the questions which has surfaced repeatedly in this and other chapters is: How can the public cope with statements that are made in mathematical or technical languages that it cannot understand? How, for instance, could the public protect itself against the dangerously misleading statements about the benefits and costs of mammography?

One answer is: You should take nothing on faith. In particular, you should avoid a misplaced faith in mathematical or other specialized languages. However, it is natural for some people to be overawed or impressed by language that is too abstruse or obscure or esoteric for them to understand. This is nothing new. It has been the main gimmick in linguistic con games that have gone on for thousands of years. Faith in prestigious, unintelligible languages has often been exploited by those who wanted to dupe or subjugate the public. Perhaps the best antidote is to look back at a historical parallel, to see how it was done centuries ago, and to see why the modern version is the same old con game. Then, perhaps, when the experts try to overwhelm you, you will respond with anger rather than with passive acceptance.

So let us go back to the late Middle Ages when the name of the game was theology rather than technology. At that time the dominant intellectual enterprise was scholastic philosophy rather than the "natural philosophy" which later became science. Then, as now, there was a constant power struggle going on. The Catholic Church was in a continuing battle with the rival religions that were then beginning to separate from the mother church. There was an even more dangerous rival--the new nationalisms that were beginning to challenge the supremacy of the Church. In the middle of this turbulence and, like good technocrats, serving whichever side paid for their services, were the experts on decision-making who were called "Jesuits" in those days.

The Jesuits achieved power and prestige by virtue of their supreme ability to "justify" any position that their masters wanted, and to do so by what sounded like impeccable arguments from scholastic philosophy. If the King or Cardinal of France wanted to wage war on the armies serving the Catholic Church, then it could be proved beyond a shadow of a doubt that this was what God in his wisdom had ordained. On the other hand, if it was expedient to stop the war, then this was the action that was unquestionably right and proper. The instrument that the Jesuits would use in either case was the obscure and highly technical language that was evolved for theological disputations. This prestigious and largely incomprehensible language overawed the opposition.

This is how faith in the powers of an incomprehensible language was used centuries ago to justify whatever decisions those in power desired. This is how this faith is exploited today. The language of scholastic philosophy is no longer prestigious. Nowadays, the language must be mathematical or at least highly technical. The manipulations required to come out with a "justification" today may differ in technique, but not in purpose. All that is really needed is that there be no direct link to reality (such as the Galilean rule). As long as "expert opinion" provides the "data," then whatever conclusion may be wanted can be delivered, then and now. Perhaps, then, it may help to think of the technocrats as "neo-Jesuits" and take what they say with the proverbial grain of salt.

The name "Jesuit" eventually became a synonym for the wily and deceptive manipulation of language. Indeed, after a while most of the parties concerned became rather fed up with the Jesuit argument. Strong efforts were made to ban the order entirely. Perhaps the same fate awaits operations research, systems analysis, and policy analysis--a fate which would be unfair in the same sense that the sins of some Jesuits were attributed to everyone in the order.

In fairness to the Jesuits, I might note that their level of linguistic sophistication (in both the good and bad senses of this word) has rarely been equaled. The disputation in public health controversies today is almost childishly simple compared to that in the theological controversies. It is not an accident that Occam's Razor, a highly sophisticated tool despite its apparent simplicity, came out of that era. Perhaps a Jesuit of those days would only be contemptuous of the crudity of the cost-benefit manipulations in the mammography issue.

PROSPECTS FOR METATECHNOLOGY

The only real answer to the growing movements to put moratoriums on dangerous lines of technological development--such as nuclear power or recombinative DNA--is to develop the technologies for the safe, effective, and economical deployment of technology which I've called "metatechnology." The limitations of metatechnology are, however, the same as those for any other technology. Any technology is a tool and can be used in ways which help or which hurt the public. The tool is neutral. It is how it is used that matters, and this will be determined by the human beings who are using the tool.

So metatechnology cannot in itself be a panacea; it is simply an instrument which could be used by the public to make decisions on the deployment of technology that would be in the public interest. Potentially, it provides a way to control runaway technologies. However, this will not be acceptable to the proponents of these technologies. They will either try to prevent the development of this decision-making instrument, or else try to take it over and use it in a neo-Jesuit mode. Metatechnology is a possibility worth considering, but it is no "pianola"--it isn't an instrument which will play itself. It has to be played in the public interest.

The chief potential of metatechnology is that it can return the decision-making on technology to the public and its representatives and thereby free them from reliance on technologists or "expert opinion." In principle, the answers to questions about the balance of benefits and costs of various plans or strategies of deployment could be obtained by what are essentially push-button operations. The machinery for estimating the doubling dose that was used for the NRC presentation can be considered a prototype for such a push-button decision-maker. In the final operation of this metatechnology, it was basically a matter of typing in a question and getting an answer at a remote terminal to a computer. Anyone can learn to do this in a few minutes.

In this way the public in a technological society can carry out its own decision-making process, one which is objective, quality-controlled, and free from the self-serving biases of "experts." In principle, the public could call upon a body of facts such as those in the Tri-State Survey, consider different deployment rules or strategies, and obtain estimates of the benefits and the costs. The balancing of benefits and costs would come from a penetrating scientific analysis incorporated in the metatechnology, and not from Mickey Mouse arithmetic. This, then, is the hope. What are the prospects that this hope will ever be realized?

From a strictly technical standpoint, the prospects of developing appropriate metatechnology for the various areas of technological decision-making are fairly bright. On the other hand, when the human factors are put into this equation, the prospects get much dimmer. The scientists and technologists who are determined to run things to suit themselves are not about to surrender their much-prized and very profitable "autonomy." They will resist to the last ditch any effort to restore authority over technological applications to the public.

Although genuine metatechnology is genuine science while decision-making by subjective opinions ("expert" or otherwise) is the antithesis of science, the battle for continued autonomy will be waged in the name of science. By calling accountability to the public "interference by politicians," the technologists may rally most of the scientific community to their cause. Hence, the prospects for restoring public control of technology are not good.

From the narrower standpoint of the technical problems of developing useful metatechnology, the prospects are good even though this is a difficult area. Perhaps in some action situations for deployment of technology the benefits and costs are so poorly estimated or so speculative that meaningful metatechnology would be impossible. An area such as recombinant-DNA research might be an example. Here, very little is really known about either the potential hazards or the possible benefits. Extreme caution is advisable in such situations. However, where there is some past experience with the technology, metatechnology would probably be technically feasible.

In at least one area, the evaluation of strategies or programs for the mass screening of women by mammography, metatechnology was developed almost all the way to the push-button stage. In this particular issue, we offered our metatechnology to the National Cancer Institute. It would have been possible, in principle, for the women who were candidates for this mass screening to have made the decision for themselves (without getting bad advice from the physicians who were pushing for this program or defending it). This never occurred. However, the findings of this metatechnology did have some impact on the eventual decision to stop the screening of the women under 50.

The metatechnology developed with Dr. Leslie Blumenson of my department dealt with both potential benefits and potential costs of various deployment strategies. The mathematical model for the growth and spread of breast cancer that provided the scientific basis for the calculations was a much deeper model than the simple one used in the NRC calculations. The model evolved over a 10-year period starting in 1966 when it was initially used to explain the results of the first and second collaborative clinical trials of breast cancer then underway. Dr. Blumenson and I subsequently used the model to predict that the third collaborative study, set up by Dr. Bernard Fisher, would produce only negative results. The prediction was fully confirmed in detail five years later.

In addition to the breast model, Dr. Blumenson developed a general theory of mass screening, another deep mathematical model. Amalgamating the two models and using general information of breast cancer to estimate the parameters, it was possible to predict the benefits from the Breast Cancer Detection Demonstration Program before any were reported. The American Cancer Society reports again confirmed the accuracy of the predictions. Dr. Blumenson's work on the screening theory won him the Snedecor Award of the American Statistical Association in 1978.

The computerized metatechnology for the mammography deployment predicted the benefits for whatever screening strategy was used. By typing in the parameters of the strategy at a terminal (e.g., age screening started or stopped, interval between mammograms, etc.), the numerical benefits would be printed out in a matter of minutes. The costs were separately calculated. The metatechnology did not produce benefit-to-cost ratios, but this would have been a simple addition to the program. All this was available at the time that the radiologists were making their crude calculations (and their arithmetic mistakes).

The National Cancer Institute was polite but uninterested in using this decision-making tool. They were already committed--despite Dr. Bailar and other critics who were part of the NCI--to the screening program. The moral is: Metatechnology has the *potential* for returning to the public and its representatives direct control over the decision-making in the deployment of technology. Unfortunately, those who favor a particular deployment (which the metatechnology may not favor) will resist all efforts to give control back to the public.

chapter 14

BIG SCIENCE VS. LITTLE SCIENCE

Science as an all-too-human enterprise:
The cover-up of radiation hazards.

SCIENCE IS A HUMAN ENTERPRISE

Modern science is a human enterprise. The human beings in this enterprise are neither supermen nor devils. They have the same strengths and weaknesses as other human beings. Indeed, for the most part, the behavior of scientists as they go about their business is not very different from the behavior of other human beings in our society. They have a narrow, self-centered view of their own enterprise and of our technological society rather than a broad panoramic view. Hence, when it comes to understanding the "big picture," the major currents and cross-currents of our technological society, scientists are just as confused as anyone else. There has been a very rapid transition in the scientific enterprise since World War II, and few scientists have been able to keep up with these sweeping changes.

What has happened to science has happened previously to other human enterprises, particularly those in the business world. A generation ago, modern science was still what economists call a "cottage industry." The basic unit of organization was a small laboratory or working group, generally an experienced senior researcher and younger assistants. With few exceptions, scientists were really "poor but honest workingmen." While the enterprise has always been dominated by a few leaders or persons with big reputations, it used to be one of the more democratic human enterprises. Then came the big money.

When billions of dollars are poured into any human enterprise, the nature and character of the enterprise is bound to change. Scientists don't like to dwell on the changes from the "good old days," often preferring to live in the past, but their enterprise has gone from a little business to Big Business in a single generation. What happens when this transition occurred in other enterprises has not been particularly pleasant or beneficial to the public. It has the same sordid patterns in the sciences. While there is no point in being sentimental about the past virtues of the scientific enterprise, there is also no point in refusing to face up to the present unpleasant realities.

It shouldn't be surprising that most members of the scientific community have a dim and distorted picture of what is going on nowadays in their own enterprise. The same would be true for most members of the business community, even though those enterprises have had three or four generations to gain a better understanding of the effects and side effects of bigness. Even today, however, relatively few persons in the business community have had first-hand experience with the decision-making processes of giant corporations. The cliches and catch-phrases popular in, say, the chamber of commerce of a moderately large city are more likely to echo the sentiments of the Horatio Alger days than the realities of the conglomerates and multinational corporations that now hold the economic power. In the scientific community there is a similar gap between the way scientists publicly speak or write about their enterprise and what actually goes on in the enterprise. The gap may be less in private conversation.

As in the business world, most persons in the scientific community have little first-hand experience with the decision-making in the multimillion-dollar phases of the enterprise. Particularly in the academic world, the more sheltered areas have not made the transition from "cottage industry" to Big Business. The scientists teach classes and supervise the research of graduate students and get modest grants. Their contacts with the Big side of the enterprise are likely to be favorable because there is a "trickle-down" flow of federal funds for meetings, publications, travel, and other fringe benefits of academic life. For a few, the "trickle-down" is more lucrative--consultantships for some influential individuals, or for persons who will testify or otherwise serve the granting agency or corporation.

For a small but powerful segment of the scientific community, however, the situation is entirely different. These persons are running the equivalent of medium-sized or large corporations. They handle multimillion-dollar or multibillion-dollar budgets. Sometimes the growth in the past generation was so rapid that the managers hardly realized what was happening. For instance, in the 1970s I suddenly realized that I was running a research business that had a half-a-million-dollar-a-year gross, although my natural inclinations are very much against the empire-building that goes on in biomedical research. This is a very small operation, relatively speaking, and the real empire builders might gross 5 or 10 or 20 times this amount.

The problems of running a Big Business, even when it is largely subsidized by the federal government, tend to be administratively similar whether the business is called "medical research" or some other name. From a management point of view a giant laboratory or institute or hospital is not so very different from a giant factory or business. So, as might be expected, the speech and behavior of the managers of Big Science is closer in linguistic pattern to that of the managers of Big Business than it is to that of their colleagues in the scientific community. Indeed, one of the concommitants of bigness in any enterprise is a steadily widening gap between those few persons at the top of the pyramid and those at the bottom. For those at the top, the original ideals of the enterprise are an encumbrance, something ignored in actual operations and used only in public relations activities.

So if concerned citizens are going to deal with members of the scientific community, they need to know whether they are dealing with "little science" or with Big Science. If they are dealing with "little scientists" they are dealing with human beings, people who may speak an unintelligible language but who can, if forced to, carry on a dialogue in ordinary English and in a sensible way. If they are dealing with Big Science, then it is no different from dealing with any other Big Business, and the name of the game is power politics. There is no way to get a genuine dialogue, although there may be passionate protestations of "concern" and various public relations gimmicks to defuse or confuse the issues.

How does Big Science operate? Perhaps the best short course on this subject is a brief history of the Sciencegate episode. This episode was mentioned in the NRC presentation as part of the background of controversy over the hazards of low-level ionizing radiation that has been going on for more than 20 years. Like its namesake, Watergate, Sciencegate is not a pretty episode, but it is one which teaches a number of important lessons.

SCIENCEGATE: THE COVER-UP

On February 8, 1978, an ex-GI who had been exposed to the fallout from the 1957 Big Smoky weapons test told his story to the House Subcommittee on Health and the Environment. Prior to the test he had been told that the fallout was "harmless"--this was the official line of the Department of Defense (DOD) and the Atomic Energy Commission (AEC). His exposure and that of thousands of other servicemen was used in a DOD-AEC publicity campaign to convince the American public that this low-level ionizing radiation was "harmless." Now, 20 years later, this ex-GI was probably dying from the delayed effects of this "harmless" fallout. He was one of the first victims of the DOD-AEC cover-up of the truth about radiation hazards that I will hereafter call "Sciencegate." But he is only the first of what will eventually be thousands of Americans who are going to sicken or die as a result of the Sciencegate cover-up.

Sciencegate began about 25 years ago when there was widespread public concern about the hazards of fallout from nuclear weapons tests. The top DOD-AEC brass were also concerned, but not about the health of the public. They were concerned because the clamor for a moratorium or a halt to weapons testing threatened to curtail these programs. Those were the Cold War days when, in the name of national security, almost anything could be done. It was not difficult to set up Sciencegate to counteract the public furor over fallout.

The DOD and AEC were the first of what are now called the Big Science federal agencies. In addition to their own enormous clout, they could count on the full support of their constituencies--the defense industries and top leaders of the scientific community who were also deeply involved in the weapons testing program. This military-industrial-scientific complex was determined to protect these programs from the clamor over fallout.

Their first step was to set the official DOD-AEC line, a line which was to hold firm for 20 years: "Low-level radiation is harmless." This wasn't true then and it isn't true now. However, in those days the scientific evidence was somewhat fragmentary. It is only comparatively recently that the serious hazards of low-level radiation have been clearly shown from data on humans exposed to radiation at this level. One important reason is that the radiation effects are insidious, and it may take 20 years before they show up.

The next step in Sciencegate was to make sure that the scientists did not deviate from the official line. This was not entirely easy because first-rate scientists are not only honest, they are also stubborn. If they found evidence of hazards, they would insist on reporting their findings. For Sciencegate to work there had to be a tough enforcement of the DOD-AEC line. The polite term is "disincentives." In plainer English, these are the actions taken to neutralize or punish any scientist who insisted on publicly reporting positive findings. In the USSR the disincentives would have been applied directly by the government, but this wouldn't work in America. In Sciencegate, the punishment of the scientists who deviated from the DOD-AEC line was carried out by the technologists who were the allies of the Big Science agencies. This scenario might be titled: "Big Science vs. little scientists."

The deliberate exposure of tens of thousands of servicemen to fallout from the weapons tests at Big Smoky and other installations was another phase of the Sciencegate cover-up. As Congressman Tim Lee Carter of Kentucky repeatedly pointed out at the hearing, of the 447 servicemen exposed to fallout at Big Smoky whose health status was known, eight had developed or died of leukemia. A simple calculation shows that at least seven of them probably got their leukemia from the fallout. Unfortunately, this is only the beginning of this man-made epidemic of cancer. One immediate effect of the hearings was that the DOD set up a hotline for servicemen exposed to fallout at nuclear tests. This is how the public learned for the first time that hundreds of these servicemen now have cancer and other diseases.

Much of the credit for bringing out the truth about Sciencegate belongs to Congressman Paul Rogers of Florida, his colleagues, and the staffers on the House subcommittee. In a series of hearings which started in January 1978 and continued for weeks, the skillful cross-examination brought to light the many pieces of the complicated Sciencegate jigsaw puzzle. Then they began to fall into place. The committee heard from the Department of Defense and from the Big Science federal agencies that are the successors of the AEC, DOE, and the NRC. Other alphabetical agencies, such as BRH (Bureau of Radiological Health), NAS (National Academy of Sciences), and NCI (National Cancer Institute), were also involved in the cover-up to some extent.

The committee also heard testimony from some of the distinguished scientists who had suffered the disincentives of the Sciencegate cover-up during the past 10 years. The committee heard how Dr. Karl Z. Morgan, who is recognized as the father of Health Physics in the United States, was stopped from presenting his findings to an international scientific meeting just before he was to speak. Nor were Nobel prizes any better protection against disincentives in the United States than in the USSR. Dr. John Gofman and Dr. Arthur Tamplin found this out. The long list of scientists who were victims of the Sciencegate cover-up continues down to the present. On February 8 and 9, 1978, the committee heard from the DOE the sordid story of how it had tried to suppress the findings of Dr. Thomas Mancuso on the hazards to workers at the Hanford nuclear reprocessing plant. On February 14, the committee heard how research funds for our studies of the hazards of diagnostic x-rays were cut off by the NCI.

Let me focus on the Mancuso story since it is a classic example of the tactics used by Big Science federal agencies to squash "little scientists" whose findings are considered a threat to technological programs or policies of the agencies. Under Subcommittee cross-examination a complicated story came to light. The technical aspects would be confusing to the public, so I will give a brief and somewhat simplified synopsis.

SCIENCEGATE: MODUS OPERANDI

By the early 1970s (when this story begins), the furor over fallout was subsiding but there was a growing furor over radiation hazards in the nuclear power cycle. Despite the growing scientific evidence against the DOD-AEC line that low-level radiation was "harmless," this was still the official position. The frequent claim that "no worker ever died from low-level radiation at nuclear installations" was the main line of defense of the nuclear industry. Sciencegate was now operating to protect civilian radiation technologies rather than military ones.

A confusing aspect of this story is that during these years the AEC became ERDA and the ERDA became DOE. The principal actors in this scenario stayed the same even though the name of the agency changed. For simplicity, the whole episode will be called a DOE cover-up. The Big Science principals are Dr. Sidney Marks, Dr. Walter Weyzen, and Dr. James Liverman. The "little science" victims are Dr. Thomas Mancuso and his eminent coauthors, Dr. Alice Stewart and Mr. George Kneale. The Mancuso team told its story at the hearing. Then the DOE team headed by Liverman came on strong to try to refute and disparage the "little scientists." However, DOE found itself on the receiving end of a devastating cross-examination by

Paul Rogers and his colleagues. In the end, the credibility of DOE was in ruins.

Ironically enough, the Sciencegate machinery that crushed Dr. Mancuso was set in motion by what turned out to be a false alarm. DOE heard that a Dr. Samuel Milham had discovered that the workers at the Hanford reprocessing plant in the state of Washington had increased risks of cancer. The DOE, always public-relations oriented, panicked. This report might threaten the official line that the nuclear radiation was "harmless" and might be ammunition for the anti-nuclear forces. Immediate steps were taken to squelch Dr. Milham. As it turned out, however, his report was to have very little public impact.

As part of the Sciencegate machinery for discrediting a positive report on radiation hazards and its author, a press release was prepared disparaging the Milham findings for release by Dr. Mancuso. At this time, Dr. Mancuso's own study of the hazards of the Hanford workers had not yet led to any positive findings. Nevertheless, as a careful scientist, Dr. Mancuso felt it was too early to issue a negative statement that could prejudice his future findings. Acting as a "little scientist," he refused to issue the press release.

This was, I think, Dr. Mancuso's "fatal mistake." In the eyes of this Big Science federal agency, Dr. Mancuso had become a potentially dangerous man. As custodian of extremely sensitive information about the radiation effects in the Hanford workers, he was a person in a key position who could no longer be trusted. Since the agency could not afford to have a scientist in such a position, the study had to be taken away from Dr. Mancuso. Thus, alarmed at the possibility that Dr. Mancuso might blow the whistle on Hanford if he did get positive results later on, the next phase of Sciencegate went into operation to kill his study.

The key to the success of Sciencegate is that it used the regular machinery of the sciences against the scientists who deviated from the official line. This is relatively easy because the peer review process which is used in the decision-making of scientific organizations, technical journals, and in the allocation of money for the support of research programs is traditionally a *closed* or secret process. A few hand-picked scientists sit in secret judgment on their colleagues. The decisions are announced anonymously, and there is little opportunity for challenge or appeal. This makes it relatively easy to rig a "peer review" to get a particular decision.

Therefore, the DOE called in six reviewers to "decide" whether or not to continue support for Dr. Mancuso at the University of Pittsburg. Four of the six recommended that the support be continued, two reviews were unfavorable. In "little science" this would have assured support for Dr. Mancuso. However, in Big Science, what the agency wants the agency gets.

Accordingly, on January 8, 1976, a DOE staffer (who subsequently left for the private sector) produced a memo recommending termination of the Mancuso contract. The recommendation was actually carried out by a second DOE staffer who replaced him. As "justification," the DOE memo cited the two unfavorable reviews, only one of which had recommended termination (and transfer to another school of public health). The four favorable reviews were not mentioned. At the time of the hearing, it was disclosed that the contract for the Hanford study that had been taken away from Dr. Mancuso had been transferred to Battelle West, a private contractor. There, it was under the administrative control of the ex-DOE staffer who had written the termination memo. This transaction was the subject of pungent comments by Chairman Rogers.

This is how Sciencegate operated to crush Dr. Mancuso. The way in which the truth finally came out is a little like a Perry Mason story. The first break came when I finally received a copy of the January 9, 1976 memo under the Freedom of Information Act. This wasn't entirely easy to get. DOE was not responsive to my original request. It took some pressure from the White House, from Senator Brooke, and from other congressmen before DOE finally sent the memo. With this memo was a letter from Dr. Walter Weyzen informing me that this was all of the information on the Mancuso decision in the DOE files. Hence, I had no idea that there had originally been six reviews. However, the Rogers Subcommittee had dug deeper.

For me, the high point of the hearing was when the Subcommittee got into the DOE review process.

> *Mr. Rogers:* Well, now what is the Committee to conclude, Dr. Liverman? You told us that your judgment was based on a staff memo which told you that the peer review didn't recommend that Dr. Mancuso continue with the study. Now we see the memo that you referred us to, and that recommendation simply is not there.
>
> What was the reason for the cancellation of the Mancuso contract?
>
> *Dr. Liverman:* I believe as we had attempted to outline for you yesterday, sir, the decision to do that was based on this memo and on the many other peer reviews, and other factors that are tied in with it.
>
> *Mr. Rogers:* Well, what peer reviews, because the majority of the peer reviewers recommended that the project continue. We went through that yesterday.
>
> *Dr. Liverman:* May I ask Dr. Weyzen to discuss the peer review with you?

After much beating about the bush about peer review, Dr. Weyzen read the "only firm recommendation." This was the recommendation to take the study away from Dr. Mancuso.

> *Mr. Rogers:* Yes. Are you saying that this is a consensus?
>
> *Dr. Weyzen:* I am not saying this is a consensus, sir.
>
> *Mr. Rogers:* This is one out of five, and the only one out of five who had suggested a change of the contractor, right?
>
> *Dr. Weyzen:* I am just trying to point out . . . .
>
> *Mr. Rogers:* Well, just answer the question--one out of five, right?
>
> *Dr. Weyzen:* One out of five.
>
> *Mr. Rogers:* That is all.

After going into the whole peer review process used to take the study away from Dr. Mancuso, Congressman Rogers summed up as follows:

Mr. Rogers: It's the most disordered, unstructured mess that I have looked into in some time. If our research programs are being carried on in this matter, where you just take a study from one scientist and give it to some other group without even knowing who the principal investigator will be or his qualifications, this is a very inefficient, poor way of managing a research program, and it is not a competent way to spend tax dollars. We are going to have to go into this in some detail, and I will ask other committees to do so as well. We may also ask the Department of Justice to look into this whole matter. Certainly you may comment.

Dr. Liverman: After your comment, what's left to say?

This and other testimony in Serial No. 95-179 gives the public perhaps the clearest insight to date into what Big Science is all about. The purposes and procedures of "little science" are distorted--often completely reversed--in the operations of the federal health and science agencies, and by the segments of the scientific community that derive their support from these agencies. It was a sordid story. However, despite these exposures, the cover-up story did not get into the media; it was business as usual for the DOE, and even though Dr. Mancuso's contract was terminated, the attacks on him have continued without letup.

Does the public have any recourse against these powerful federal agencies? Reform is probably impossible for entrenched bureaucrats. A Proposition 13 approach would seem to be the only hope. Possibly the only way to deal with the DOE is to abolish the department.

SCIENCEGATE: THE CONSEQUENCES

The Sciencegate cover-up engineered by THE DOE had apparently nailed the lid down on any report to the public of the dangers to the Hanford workers. Actually, the cover-up backfired. The Mancuso report on the low-level radiation hazards ended up with worldwide media coverage, in part because of the DOE attempt at suppression.

To many people, the most surprising part of Sciencegate may be that it was possible for this suppression of science to go on for 20 years in an open society such as the United States. What must be remembered, however, is that inside our open society there are some tightly closed subsocieties. The Big Science federal agencies are not much more open than the CIA or FBI. In particular, the "civilian" DOE or NRC are still involved with nuclear weapons and still operate under a National Security blanket.

Unless one has had first-hand experience with Big Science agencies such as the DOE and NRC, it is hard to appreciate that they are private worlds, closed societies in the midst of our open society. With such powerful constituencies as the nuclear industry or the defense contractors, the agencies feel immune to public pressure from Congress or even from the White House. Congressman Rogers could call the DOE's actions in the Mancuso affair "the worst mess" he had ever seen in research, but Dr. James Liverman was hardly contrite. He later went on television in a further effort to discredit the work of Dr. Mancuso. Nor did Dr. Sidney Marks seem in any way disconcerted or apologetic for what he had done.

A "little scientist" or an ordinary citizen has no defense against the actions of Big Science agencies. Until I sent a copy of the memo that I secured under the Freedom of Information Act to Dr. Mancuso, he had no idea of how or why his contract had been terminated. Even now there is nothing he can do about it, and he has no funds to continue his important research on the hazards of low-level ionizing radiation. The Big Science agencies and their allies have the power to take away a scientist's reputation and livelihood and have no compunctions about doing so. This is why the list of American scientists who have been punished for their positive findings on radiation hazards is longer and has more first-rate scientists on it than the list of Soviet scientists who were punished for their genetic findings during the Lysenko era in Stalinist Russia. While the official line that radiation is "harmless" is not an ideological one, it is far more dangerous to the public health than any political slogan.

The real horror of Sciencegate isn't what it has done to American science, but what it has done to the health of all Americans. Disastrous consequences of Sciencegate are still to come. Because it was official that low-level ionizing radiation was "harmless," tens of thousands of nuclear workers have been exposed to dosages which will produce cancers and other diseases in years to come. Again, because of the DOD-AEC doctrine, thousands of radiologists treated millions of Americans with "harmless" doses of x-ray for trivial conditions such as acne or for self-correcting problems such as postpartum mastitis. The NCI has finally issued a pamphlet urging the unknown millions who got this unnecessary and dangerous x-ray exposure to come in now for a check-up for thyroid cancer. Eventually this check-up will have to include breast cancer, lung cancer, and other cancers.

Just as thousands of servicemen were exposed to radiation from fallout because of the myth that it was "harmless" and 20 years later are getting fatal diseases, the American Cancer Society marched a quarter of a million American women through radiation from mammography that was also supposed to be "harmless." In 20 years, the women under 50 will be getting breast cancer and other diseases from this unnecessary and dangerous radiation. At present, many Americans are getting CAT scans which have sufficient x-ray dosage to double their risks of leukemia. They too will be victims of the Sciencegate cover-up.

For the past 20 years radiation technology has been promoted as "harmless" and has often been used in a mindless way by those who promoted it. The price of this in genetic damage to the persons exposed and to their children is a price that must be paid in suffering and death.

chapter 15

HOW TO BECOME A MILLIONAIRE

How application of the strategies in this book have actually saved lives.

AN ANTIDOTE?

The title of this chapter may seem out of place in a survival manual for thinking men and women of a technological society. The tone of this chapter may also seem a bit out of character--overly optimistic and boastful, rather than pessimistic and realistic. But anyone who tries to apply some of the suggestions of this book in an effort to save his or her own life from any specific misuse of new technologies will discover that this is a very slow and discouraging process. It is easy to fall into the attitude which the public relations campaigns of the major chemical polluters deliberately try to foster: We live in a sea of carcinogens and other hazards, so it is really hopeless to do anything to reduce exposures to the dangerous chemicals that are being dumped in the environment. Give up and learn to love swimming around in the gunk.

So the purpose of this chapter is to try to provide an antidote to these deliberate efforts to produce despondency, despair, and, finally, complete apathy. In the past year or two, this croaking chorus has been joined by public health bureaucrats and professors of preventive medicine who want to justify their own failure to act in the public interest by pleading that nothing can be done to protect the public against this deluge of dangerous chemical products. An alternative theme is that it is all the fault of the public and not the polluters, that the public can change its lifestyle to protect itself if it wishes. Those who want to poison or be poisoned should be allowed to "do their own thing without government interference." This latter point is what the industry-orchestrated chorus is all about: Nothing should be done to restrict uses of technology (or the opportunities for a fast buck).

However, even without the public relations campaigns by Mobil and Dow and other major polluters, it is often difficult to avoid falling into an apathetic posture. It is so hard to get bureaucrats at any level to listen to your problems and do anything about them, to get politicians to keep their campaign promises, to get support from fellow citizens when pressure is needed, and, indeed, to get anything done in these complex action situations. It is hard to sustain determination and activism over the period of many years that is likely to be necessary to get anything effective done. It is all too easy to get discouraged, to give up, and to abandon all hope of saving your life and the lives of others.

So what I would like to do in this chapter is to suggest that the situation is not quite as hopeless as it often looks. One individual with a bit of luck and a lot of help from others *can* achieve results. So, here then is a case history report on what was accomplished by one not-exceptionally-talented individual.

A STRONG AMBITION

When I moved to New York City and Cornell University Medical College in 1952, I left the protective academic environment of Johns Hopkins. Without intending to do so, I went "into business for myself." Although I had thought that my salary and the costs of my small statistical unit would be covered by the Rockefeller Foundation and the university, I soon discovered that I was really expected to go out and get enough dollars to pay my own way.

In the biomedical research area, this is the customary state of affairs; the investigators are really entrepreneurs who sometimes get little more than a mail drop from their institutions. They have to go out and get their own research money, usually from the federal government. They may have to raise most of their own salary, even in "tenured" positions. So if biomedical researchers sometimes are afraid to do "controversial" research (and this covers almost all research relevant to human beings and their problems), it is often because they are very insecure--particularly if they are paramedicals in a medical environment.

In my case, this insecurity had two very favorable consequences. It forced me to work very closely with two programs, one at Cornell and one at the Sloan Kettering Institute. These programs each contributed a few thousand dollars to the costs of my statistics section and enabled me to balance my budget. But even more important, these programs were directly involved with major public health problems and led me to nurture a strange ambition--the ambition to become a millionaire.

The more I worked with these programs, the more I was impressed with their potential for saving human lives--not in the abstract sense of some remote future time, but in the real sense of saving actual persons here and now.

It was at that time, early in the 1950s, that I got the ambition to become a millionaire--not in dollars, but in human lives. I wanted to be able to assert *and to document* the fact that my work (together with others, of course) had saved 1 million humans from dying prematurely from highway accidents, cancer, and other causes.

From what we had accomplished by, say, 1955, the prospects for my becoming a millionaire looked very good. Even in retrospect, it is clear that if what we had learned about how to save your life and the lives of others had been translated into effective public health action, I would probably be a multimillionaire by now. At that time I was too young and innocent to anticipate that there would be bitter, last-ditch resistance to these efforts to protect your health and safety.

Despite many failures and only a few successes, the rough tally that will be presented comes respectably close to the target of 1 million lives saved--just how close I'll leave it to you to judge. I don't want to get into a numbers game here. The real point is not whether I ever will save 1 million lives or not, but that there was and still is a huge potential for saving lives by primary prevention. The potential was there in 1955, and it is there today. It can be realized if you don't give up hope and if you don't let yourself be brainwashed into learning to love "swimming around in the gunk."

Perhaps you feel that my ambition to save 1 million lives converts a humane effort into inhumane statistics. In extenuation, I can only say that I am a statistician. My business is the objective measurement of achievement. If 1 million is an abstract number, the persons whose lives have been saved have been real flesh-and-blood human beings.

One other point: In addition to showing that I have saved many lives and could have saved many more, I want to show you *how* this was done. You will see that there was no mystery or magic to this. There was no multimillion-dollar high technology instrumentation, no huge research staffs and budgets, and none of the costly trappings of Big Science. It was only a matter of getting the facts and then using the scientific strategies described earlier to draw sensible conclusions about how to save lives. It was common sense reduced to calculation that did the job.

A FACE IN THE SEAT

It all started with a face in the back of a seat. I had just come up to Cornell University Medical College in New York City, and my office was next to an oddly furnished space. There was ordinary office equipment in this area plus what seemed to be a large collection of smashed, twisted, and useless junk. Later, I learned that this junk was actually equipment that had been taken from light aircraft crashes. It was used in crash-injury research.

One of the items was an aluminum seat back with the imprint of a face in it. This impression had been made by the passenger in the seat behind who had hit the back of the seat during the crash.

Incredibly enough, this passenger who left this imprint in metal actually got off with minor injuries and a headache. It was a preventable death that had been prevented by good seat-back design.

By the time I arrived at Cornell in 1952, this aircraft safety program, founded by Hugh DeHaven, had already achieved some remarkable results. I was amazed to learn that you had a better chance of getting out alive in a well-designed light plane crashing at 150 miles an hour than in an automobile crashing at 30 miles an hour. It occurred to me immediately--as it would have occurred to you--that if you can save hundreds of lives by improving the crashworthiness of light planes, you can save tens of thousands of lives by doing the same thing with cars. I used to have some hot arguments with my next-door neighbors in my efforts to convince them that they should shift their target. Eventually, the project did become the Auto Crash Injury Research Program of Cornell University (ACIR).

At the beginning, the major automobile companies provided the funds for the ACIR program, and everyone got along just fine. John Moore provided the administrative push and Boris Tourin did most of the actual work, while I provided the ideas and statistical know-how. The basic approach was really very simple. We developed a better form for reporting highway accidents. The state troopers filled out this form at the scene of the crash. Later, this data was supplemented by medical information from the doctors who treated the injured occupants. From this report we could try to analyze the sequence of events leading up to the death and injury. Somewhat unexpectedly, we hit pay dirt almost immediately. However, it took us some time to realize it.

An unsung hero of this story is an Indiana state trooper, Sergeant Paul, who had noticed that when occupants were thrown out of a car in a highway accident they usually were badly hurt or killed. At the time, however, the safety experts "knew" that an occupant was better off to be ejected. There were even "safety devices" that facilitated the ejection.

To settle the issue, we put questions about doors coming open and ejection of the occupant in the questionnaire. Back in those days we could ask any questions we liked. Federal regulations on research involving human subjects and the malpractice issue were not an obstacle then. So within about a year we had about 100 completed case histories and some definite findings to report.

We hadn't expected to be able to say very much in our first presentation to Detroit because we thought that a study of thousands of accidents would be needed. But Sergeant Paul was so clearly right in his observations about ejection that it was already clearly identified as a killer. What was evident, even in this small original sample, was that most of the ejection was occurring in rollover accidents and through doors which had been popped open during the roll. Indeed, elementary physics supplemented by experience with aircraft accidents immediately suggested the sequence of events that led to death from ejection.

In brief, the frame of the car would be distorted in the rollover, and this would release the latches on the car door. Centrifugal force would then open up the door and throw the occupant out of the car. The occupant would be killed or injured unless he or she was very lucky in landing.

Once the statistical analysis of the accident reports led to a coherent picture of what was going on in the ejection accidents, there were several possibilities for interrupting this fatal chain of events. One was to put door locks on cars which would hold even in rollover accidents. Another possibility, naturally suggested by experience in aircraft accidents, was to restrain the occupant in the seat by a seat belt. However, new devices may not work the way you think they will. It is always necessary to evaluate any protective measures very carefully. *Intervention in any health issue always involves the risk that there will be side effects or unexpected chains-of-events that are themselves hazardous.* Therefore, we looked to see if occupants were hurt by doors which didn't pop open and, later, to see if the seat belts were producing injuries. You have to be careful.

Seat belts, and some of the other methods for "delethalization" of highway accidents that came out of the early work by ACIR, were one of the relatively few major advances toward saving your life that have been made in the past generation. What I want to emphasize is that *there is no magic or mystery* about how this scientific approach to a major health problem worked. We went out in the field to get the facts--in this case, out on the highways where the events actually were happening. We used modern biostatistical technology for drawing sound conclusions from the factual evidence to gain an understanding of what was happening on the highways. When we had a good idea of what to do about a hazard, there was an interlude of engineering and laboratory development. It was necessary, for instance, to devise new door locks and adapt safety belts to automobiles. Then it was back to the field to see if the devices worked and to get new life-saving ideas. In this whole cycle, there is nothing which cannot be explained in plain English and in common sense terms.

What I've just said is in flat contradiction to what you have always heard from the mass media about scientific research. You have been told that research is something very marvelous and mysterious, and that it involves all sorts of things that you would not or could not understand. This mystery and mumbo-jumbo about science--this mystique or ideology--is just the opposite of what science is really all about.

COUNTING THE SCORE

How shall we score the results of this ACIR experience? In July 1960, I published a paper in *Public Health Reports* on "How to Cut the Highway Toll in Half in the Next Ten Years." It presented a specified series of preventive measures, and using ACIR facts and simple biostatistical methods, showed how the traffic toll could be cut in half if their measures were implemented. Apart from seat belts and a few other "delethalization" design changes, however, the actions were not taken and the highway toll was not, in fact, visibly reduced. In theory, the number of lives that could have been saved in highway accidents over the past 20 years or so might have been over 700,000 (at, say, 35,000 a year). This one study might almost have made me a millionaire. But, unfortunately, "might-have-beens" don't count.

There was some definite savings in lives from the seat belts and other design changes. An ACIR study carried out later estimated the savings in lives from the seat belts at 6000 per year. Over 20 years this comes to 120,000 lives. While this is a substantial achievement in its own right, it is disappointing in terms of "might-have-beens." Or were the potential gains an illusion? We'll come back to this question in a later section.

Why wasn't more of the potential achieved? Perhaps my main failure was in communicating our scientific findings to the general public. I had published these findings where they could be read by scientists concerned with highway accidents. In the traditional view, I had done my job. After this, it was up to "someone else"--federal agencies, the public health community, the public or its representatives--to translate these life-saving ideas into actual public health measures. Unfortunately, there wasn't "someone else" to do the job, so the job wasn't done.

Even such an innocuous action as putting a scientific paper in literature was not without risk. Detroit tried to get me fired (and eventually did end my consultation with ACIR). If I had been at Cornell when the paper came out, I would have been in real trouble. However, by then I had moved to Roswell Park Memorial Institute (RPMI) and was in cancer research. My new boss was Dr. George Moore, Director of RPMI. He called me up to tell me about the pressure from Detroit, and he then told me to forget it. That was the end of it. All my career I have been just plain lucky this way.

Why didn't I go directly to the public with the truth about highway accidents--wasn't an opportunity to cut the highway toll in half going to be big news? You might think so. However, the media didn't. Perhaps my mistake was aiming for the "elite" media--the *New York Times* or the *Scientific American*. This was a natural mistake since you would suppose that there might be a better chance of publishing material that is in the public interest in such places. It took some years of additional experience to learn that the reverse is true.

FOLLOW-UP ON HIGHWAY SAFETY

The amount of fraud and fakery that now goes on in the name of science is staggering. I first became aware of this from my ACIR experience. At the time I thought the situation was peculiar to highway accidents, but I have seen the same thing in a dozen health areas, including several of the currently fashionable areas of cancer research.

At ACIR, what bothered me was this: We had reached a whole series of important findings on how to save lives in auto crashes within the first three years. *Now why hadn't this all been done long before?* Why, for instance, had the argument about whether an occupant was safer if ejected gone on for years when even a little good data could have settled this point immediately? We hadn't done anything particularly exceptional or brilliant. All of it could have been done 20 years earlier--but wasn't. Why wasn't it?

The answer was that, for all the endless talk about highway safety, *no one had really bothered to look at the problem in a genuinely scientific way*. No one had taken the trouble to build up a good factual base for studying highway accidents. No one had applied modern statistical technology for finding the truth from a body of evidence. In other words, practically all of the so-called "safety research" that had been going on for so many years was nothing but a fraud. Similarly, practically all of the so-called "safety education" was nothing more than a parroting of Detroit's ideology on this subject. We had discovered these important results so quickly because we happened to be the first ones who really took the trouble to look.

History repeats itself. Health physicists and "health educators" have been assuring the public for years that low levels of radiation are perfectly safe. *No hazards had been found because no one really took the trouble to go into the field and look*. The "experts" merely recited the ideology of the Atomic Energy Commission on this subject. Once scientists took the trouble to look to get good data, to use powerful new statistical technology, and make an honest report of what they found, all sorts of hazards have turned up. My staff and I are not so bright that we can scoop our colleagues time after time on these health issues. It is simply that we have taken the trouble to look carefully in the place where the hazard actually occurs--in the field.

You might innocently suppose that when ACIR began to make real progress toward saving lives--in an area where no progress had been made for half a century--there would be appreciation or approval from those who had so loudly proclaimed their interest in highway safety. In fact, ACIR became very unpopular all around.

For one thing, ACIR had become a dangerous new rival to the "highway safety establishment" both for funds and for public attention. The persons doing the ineffectual traditional "research" all hated ACIR. The National Safety Council, which had been pushing the Detroit ideology for years (the "nut-behind-the-wheel" or "the-public-is-the-guilty-party" line), the "safety experts" and "safety educators" who had echoed the Detroit line, and the rest of the establishment pooh-poohed our findings. Even the automobile manufacturers were beginning to wonder if they had built a Frankenstein monster in ACIR. They had been delighted with the publicity bonanza they got from their ACIR support. Now, however, ACIR was beginning to tell the auto stylists and designers what to do. This was intolerable to Detroit.

Following up this early ACIR success to see what happened after this promising start, what happened next was both good and bad. On the favorable side, the evidence began coming in that the early delethalization efforts were beginning to have a modest payoff in human lives.

While we were off to a good start, some other steps were needed to make a major reduction in the highway toll. In my 1960 paper, I had listed these steps. Some of these were long overdue--standardization of the bumper heights on all vehicles, for example.

Others were a little more subtle and involved, e.g., doing something about the mix of large and small vehicles on the highway. We soon found that it was particularly hazardous for occupants of lighter and smaller vehicles when the accident involved a much larger and heavier vehicle. There was one other very important factor in a crash that was already obvious in the first 100 cases, and this was the speed at impact.

SPEED KILLS

Then and now, a very substantial fraction of all fatal accidents occur at very high speeds--speeds well above the legal limits. However, only a relatively small proportion of the cars do much sustained travel at such speeds. Speed control was, therefore, a key proposal in my paper. This was to be achieved by design of the cars--by control of horsepower, for instance--rather than law enforcement. Since then I have written a series of reports showing the importance of this step.

During the 1950s, there had been a gentleman's agreement between ACIR and the automobile industry that the dirty word "speed" was not going to be mentioned. The industry has always been hypersensitive to any talk of speed control. It was an article of faith in Detroit's ideology that "speed sells cars" (and also that "safety is poison"). ACIR took pains not to upset the industry by using the speed issue. Moreover, there was the general feeling that talk about speed detracted from the delethalization theme. However, I have never felt bound by any gentleman's agreement when such an agreement was against the public interest (though my attitude doesn't win me any popularity contests in the health establishment).

Although all the "safety experts" and automobile engineers often spouted the industry ideology about speed providing an extra margin of safety, Detroit has always known that the excessive speeds they were engineering into cars were actually killing their customers. In the "honeymoon" days of ACIR and in an unprecedented burst of public spirit, the industry actually set up voluntary agreements to restrict horsepower and to play down speed. For instance, no company was supposed to sponsor racing or to race cars in its name.

Unfortunately, these agreements within industries only seem to last as long as there are public outcries to be stilled or publicity to be gained. During the 1960s, the industry went all out for a horsepower race. This not only wiped out the gains we had begun to make, but actually produced a more hazardous situation on our highways than ever before. I have seen this happen in two or three different health issues--early gains wiped out later by arbitrary actions of industry. It seems to be something of a pattern. It is as if, having done a good deed, the makers of hazardous products think they can get away with anything.

Now let me introduce another actor in this drama. The 1960s saw the appearance of several federal agencies whose specific mission was to protect your health and safety. Many people hoped that they would be the answer to the problem. By and large these agencies have been a great disappointment.

The Highway Traffic Safety Administration has taken almost no effective action. It gets publicity for noisy, costly, and ineffective measures "cracking down" on Detroit. The recall of automobiles, for example, has made a lot of noise. However, I have seen little evidence that it has saved very many lives.

These federal agencies sometimes seem completely helpless. Once I had almost talked Bill Haddon (then the top administrator in the federal highway safety agency) into trying to get an upper limit on design speed as a matter of principle. The limit he suggested was 100 miles an hour. However, the agency couldn't put this limit through against solid industry opposition!

Until the 55-mile-per-hour speed limit was imposed as a conservation measure during the "energy crisis," there was no good way to show that my predictions were correct. But when the speed limit was still in the talking stage, I wrote a letter to the *New York Times* predicting at least a 25 percent reduction in deaths in auto crashes if the limit went into effect and if it was enforced. I based this prediction on some old ACIR data I had in my files. This prediction was subsequently confirmed by state and national figures. This shows that if we had effectively used what we had learned from our scientific studies by 1960, we could have done even better than cutting the highway toll in half!

LOBBYING CAN BE HAZARDOUS TO YOUR HEALTH

As might be expected, Detroit and the big oil companies and the others in the speed lobby were not at all happy with the 55-mile limit. In their advocacy advertisements, oil companies may sound gung ho for conservation. What they *do* when they lobby in Congress or at the pump is another story. So the attack on the 55-mile-per-hour limit came from several different directions, and it is interesting to see how it developed.

The auto and oil companies support a lobbying group which is supposed to be a spokesman for "Highway Users" such as you and me. This lobby went into furious action. Despite the plain facts that for the first time in a generation there was a dramatic drop in death rates as a result of the 55-mile limit, the lobby tried to muddy the waters.

Another attack on the 55-mile-per-hour limit appeared in what, at first sight, might seem to be a strange place. It surfaced as an editorial in the *Journal of the American Medical Association*. Written by a "safety educator," it contained the usual double-talk that appears in so-called "statistical controversies" on health issues.

What was especially worth noting about the *JAMA* attack on the 55-mile-per-hour speed limit was that it occurred *after* there was massive and indisputable factual evidence that the speed limit had produced a major reduction in deaths and death rates on the highway. Although the data were overwhelming, particularly the evidence from controlled access highways, the *JAMA* editorial attempted to generate a "statistical controversy."

All this was reminiscent of the cigarette-cancer controversy. It was also one of the tactics in the efforts to keep thalidomide on the market. When the horrors of thalidomide first came to light, *Science* ran an editorial saying that the thalidomide issue could *only* be resolved by further animal studies because of supposed "statistical" difficulties with the German data. It is greatly to the credit of the woman who wrote this editorial that she later retracted it. Few people who get into these health "controversies" on the wrong side ever admit their mistakes.

At this point, I wrote a hot letter to *JAMA* pointing out in no uncertain terms that the editorial was atrocious. I suggested that this anti-public stance might boomerang into public support for socialized medicine. This letter wasn't published, but I did get a reply which promised that the editorial theme would not be repeated. I did some more letter writing to the media and to congressmen. Not long after, the steam went out of this lobbying campaign. Even if I can claim no credit for the 55-mile-per-hour limit, I did do something to help keep it in effect. My efforts had enough impact for a lobbyist for the "Highway Users" to put a letter in a Buffalo newspaper denouncing my "unfairness" to these lobbyists! Not even lobbyists like to be seen as "bad guys" by the public.

From the previous reckoning, the seat belts and other safety hardware that came in with the ACIR program have probably saved about 125,000 lives. The reduction in the number of cars traveling at extreme speeds cut the highway toll, as predicted, by about 25 percent. Some of the other recommendations in my 1969 paper, such as bumper regulations, have also been partially implemented. Despite constant pressure from trucking interests and others, it looks as if the reductions in speed and horsepower are going to continue.

The main savings in lives here date from the oil embargo, and so they have been realized for about five years. While an exact figure is not so easy to estimate, perhaps 20,000 lives a year for five years would be a reasonable guess. Adding this to the previous total would bring the saving in lives to about 250,000. This doesn't make me a millionaire (though there should be further savings in the future), but it does put me well on the way to my original goal.

WILL I EVER BE A MILLIONAIRE?

While again it is hard to come up with accurate estimates (and it isn't really important to do so), the number of lives saved as a consequence of research started at Sloan Kettering Institute with Dr. Ernst Wynder in the early 1950s, is of the same order of magnitude as the saving in highway accidents. However, the cigarette-cancer story and the development of low-tar cigarettes is too long to go into here.

Although my efforts in the cigarette-cancer issue seemed completely frustrated and frustrating for most of the 30 years that I've spent on this problem, there has eventually been a big payoff. It has come in an unexpected, but not entirely unpredictable, way. There have been at least three "tar derbies" which have driven down the nicotine and tar in most American cigarettes. Originally, the average amount of tar in standard cigarettes was about 35 mg. With the switch from standard to "king size" and then to "100s," the tar levels were headed in the wrong direction--the levels went up. However, later on, the "tar derbies" where cigarette companies competed on a low-tar basis in this highly competitive market drove the level back down. At first, it was driven down to 25 mg in the filter cigarettes. This was some reduction from the original level of 35 mg, enough to show statistically that there was a reduction in the risk of lung cancer for persons who switched to filter cigarettes. Later we could show a reduction in the risk of heart disease in the persons who switched. This was the focus of my 1975 paper, "Less Harmful Ways of Smoking," presented at the Third World Conference on Smoking and Health.

In this paper, I predicted reductions in the risks of heart disease from the new "tar derbies" and urged a push from the American Cancer Society and the National Cancer Institute in this direction.

These findings--about the only encouraging new data presented at the entire meeting--were studiously ignored by the anti-smoking establishment. They were determined to stop everyone from smoking (but not getting very far). I have always felt it was a better public health strategy to take the hazard out of the cigarettes than to take the cigarettes away from the smokers.

As I was writing this section, a United Press International report headlined "Deaths from Heart Disease" decline appeared on October 26, 1978. The story noted a drop of 18,000 deaths from 1976 to the lowest total since 1963, "despite an ever-aging and ever-increasing population." Dr. Robert Levy of the National Heart, Lung, and Blood Institute is quoted as saying that "if the heart disease rate of 1968 had continued into 1976, 164,000 more Americans would have died of heart attacks two years ago."

As the story indicates, the National Institutes of Health doesn't know how much of the reduction can be attributed to medical intervention and how much to primary prevention. It doesn't know how much is due to smoking and how much to diet and lack of exercise. Washington doesn't know because it doesn't monitor the death rates carefully or scientifically.

Although the vital statistics rates are potentially an invaluable source of information about what, if anything, federal research programs in the health area are accomplishing, they are not effectively used for this purpose. The agencies play it safe. Most of the statistics reported are little more than counts and death rates. It would be relatively cheap and easy to see what is happening to the death rates, and also *why* it is happening. Unfortunately, NIH has little interest in monitoring research productivity, the health industry, or its own operations. It prefers to let the public buy a pig in a poke for its tax dollars for medical research.

The trends to lower tar cigarettes are certainly the main factor in this decline, although it is not easy to say exactly how many of the 164,000 Americans who didn't die had saved their own lives by switching to lower tar cigarettes. It is probably safe to say at least half and possibly as much as two-thirds of the total is attributable to cigarettes as the primary factor.

In general, more than one factor might be involved in the change in risk, but these other factors often act jointly with cigarette smoking. For example, the excess risks of lung cancer from exposure to asbestos alone is small compared to the excess risk for a cigarette smoker with exposure to asbestos. Similarly, in heart disease there are other factors besides smoking. However, without the constant reduction in tar levels they probably would not have shown up in the death rates.

It is hard to say exactly when the first effects of the "tar derbies" occurred or how great the effect was each year. For rough calculations, we might take an average saving of 50,000 a year for 10 years. This would come to about half a million lives.

Adding up the lives saved in highway accidents and from development of a less hazardous cigarette brings me more than halfway to my goal of a million lives. Again, the gains from the low-tar cigarettes are only a small part of the "might-have-beens." The last-ditch resistance of the tobacco industry plus the equally obstructive efforts of the American Cancer Society and the anti-smoking establishment delayed the development of large-scale use of the less hazardous cigarettes for 10 to 20 years. If the switch to low-tar cigarettes had taken place earlier, I might have been a millionaire by now from this one line of research.

Let me draw one moral from this story and also pose one rather baffling question. The moral is that serendipity is something a public health scientist needs badly. Here, my main focus was on protecting people from cancer and only in a secondary sense on protecting them from heart disease. Yet the clear results are for heart disease, not cancer.

The puzzling question is: Why hasn't there been a drop in lung cancer death rates? Part of the answer is suggested by the new scientific facts on the co-occurrence of disease (including those in the NRC presentation). Heart disease comes earlier than cancer.

The smokers who don't die from heart disease may have a paradoxically greater chance of living to get lung cancer. But this can't be the whole story.

For some time, I have been inclining toward the view that there are other factors--asbestos exposure, for one--which are now "taking over" from cigarettes in the lung cancer rates. Most of the factors, such as occupational exposures, involve co-carcinogens that act jointly with cigarette smoking. Even at lower tar levels for cigarettes, the joint action keeps the risk of lung cancer high. We are now seeing the effects, both on men and women, of World War II changes in smoking habits, occupations, and environmental pollution. As one cancer hazard is controlled, the other hazards are getting worse, and the cancer death rates show little improvement.

PREVENTION VS. CURE

Without trying to go into a detailed accounting, I might note that our other research in primary prevention has saved a few thousand lives here and a few thousand there. I am still hoping to save substantial numbers of lives with our radiation research, although the payoff to date may not be large. In addition to prevention, there are also some lives that have been saved by our research on the treatment of cancer.

I mention this therapeutic life-saving mainly to contrast the potential savings from prevention with these savings from therapy. The first controlled clinical trials on the treatment of childhood leukemia were set up, at my suggestion, at Sloan Kettering about the same time as the accident and cigarette-cancer studies. Out of this work eventually came one of the very few real advances in the treatment of cancer in the past generation. For the first 20 years, the chemotherapy of childhood leukemia prolonged life for a few months or years, but there were few, if any, cures. Now it looks as if there may be some cures. However, we are talking about saving a few hundred lives at the most.

One other area of the therapy of cancer where some progress has been made is breast cancer. The first controlled clinical trials for breast cancer were set up in the Biostatistics Department at RPMI. These achieved the first advance in the treatment of breast cancer for 50 years. Again, the original effect was mainly to delay the progression of the disease rather than to cure it. By now, however, it is possible that some cures may be achieved with the chemotherapies for breast cancer that have evolved from our work. Again, a few hundred lives may have been saved by these new treatments.

These lives saved through therapeutic advance are as important as the lives saved through primary prevention, but they are not any more important. Of the billions of dollars spent for cancer research by the National Cancer Institute and for other disease research by the National Institutes of Health, less than 5 percent has gone for research that can honestly be called primary prevention. The NCI's Division of Cancer Cause and Prevention, for instance, has spent most of its money on virology (an impossible "cancer vaccine") or on "basic science" (laboratory studies which have told us very little that is useful about human cancer).

Nearly all of the rest of the NCI money went for "treatment research." However, after 25 years of claims and headlines, we still cannot document the cure of a single patient with generalized disease for any of the major primary sites of cancer. For cancer at this late stage, there is no more evidence of cures from standard modalities of surgery, radiation, or chemotherapy than for laetrile or faith healing.

This waste of the public's money will go on as long as the "peer review" of the National Institutes of Health is dominated by physicians and basic scientists from the elitist medical schools. For the past generation, these people have felt that "research" is a word which can be used in Congress as an acceptable euphemism for "subsidy" of the medical schools. The much-praised "peer review" is simply a device for allowing an in-group to pass out money to itself. Most of the "peer reviewers" on the decision-making committees (not necessarily low-level "site visitors") are simply administrators. They generally have little competence or even interest in medical research. They are there to get the money for their institution. This is the only thing that they do very effectively.

There are two points which might be made from this contrast of primary prevention and therapeutic strategies for saving lives. The first point is that it is still true that an ounce of prevention is worth a pound of cure. In terms of saving lives, a given investment of resources or scientific effort in primary prevention (an ounce) can produce more than 16 times this investment (a pound) in therapeutic research. In the accounting of lives saved, the hundreds of lives that may have been saved by the advances in the treatment of leukemia and breast cancer are far fewer than the hundreds of thousands of lives saved by primary prevention.

The second point is that saving lives doesn't pay--at least not for the researcher who saves these lives. Our program for Biometric Research on Cancer Epidemiology underwent National Cancer Institute "peer review" by clinical panels consisting almost entirely of physicians. The radiologist who came to the "site visit," and succeeded in killing our proposal, was opposed to our efforts to protect women from the ACS-NCI program for mass screening by mammography, and seemed concerned with protecting his fellow mammographers. In general, doctors disparage and denigrate efforts at primary prevention. They want the health industry to keep its monopoly on the "life-saving business," although it has a poor record. Until the decisions on funding research are made by persons who are interested in saving your life rather than protecting professional interests, you can't expect that very much medical research will be done that will be for your benefit.

THE WINE OF ASTONISHMENT

Trying to save your own life or that of others can be such a long, drawn-out struggle that it is emotionally exhausting. Sometimes the spirit sags. Any source of hope or encouragement can help to keep a person going. My private ambition to become a millionaire in lives saved has helped me over some bad times. It led me, for instance, to agree to go to Washington recently to meet with a General Accounting Office group to report on our low-level radiation research. GAO is supposed to be the congressional watchdog and research unit. GAO recommendations are supposed to guide legislation and provide some oversight of the federal agencies that aren't doing their jobs.

When I got down to Washington and checked in at the hotel, I got some bad news--a confirmation of what I had been previously warned would happen. I was planning to criticize NRC, DOE, and other agencies for failure to take effective action on radiation hazards. However, the consultants for GAO that I would meet the next day also happened to be consultants to the very units that I was going to criticize. This kind of "peer review" by allies of the nuclear industry or health industry is all too common with regulatory or even "watchdog" federal agencies. I had been invited to a poker game where the cards were going to be stacked against me. It was a futile trip, and I was discouraged.

There happened to be a Gideon bible open in the room. Although I'm not a dedicated reader of this particular literature, I opened the pages at random, closed my eyes, and put down my finger on a passage. Would it give any portent of what would happen when I faced my adversaries the next day? I was delighted by what I read and stopped worrying. My finger was on the third verse of Psalm 60:

> 3. Thou hast showed thy people hard things; Thou has made us to drink the wine of astonishment.

If I were looking for an epitaph, I could not hope for a better one.

BIBLIOGRAPHY

NRC REFERENCES

Note: The NRC presentation did not have the usual bibliography. Instead, the following list of references was published in advance in the Federal Register Notice.

Bross, I. D. J., Hazards to Persons Exposed to Ionizing Radiation (and to their Children) from Dosages Currently Permitted by the Nuclear Regulatory Commission. For presentation at a U.S. NRC public meeting, April 7, 1978.

Bross, I. D. J., and Natarajan, N., Genetic damage from diagnostic radiation. *Journal of the American Medical Association, 237*(2): 2399-2401 (1977).

Bross, I. D. J., Right answers from wrong assumptions. *Preventive Medicine, 5*:203-206 (1976).

Bross, I. D. J., and Blumenson, L. E., Screening random asymptomatic women under 50 by annual mammographies: Does it make sense? *Journal of Surgical Oncology, 8*:437-445 (1976).

Bross, I. D. J., Scientific strategies in human affairs: Explaining statistics to the public. *American Statistician, 30*(4):171-175 (1976).

Blumenson, L. E., A comprehensive modeling procedure for the human granulopoietic system: Detailed description and application to cancer chemotherapy. *Mathematical Biosciences, 26*:217-239 (1975).

Bross, I. D. J., Strategies of Science in human affairs: How to assure the quality of scientific statements. *Clinical Pharmacology and Therapeutics, 15*(6):543-550 (1974).

Bross, I. D. J. , and Natarajan, N., Risk of leukemia in susceptible children exposed to preconception, in utero and postnatal radiation. *Preventive Medicine, 3*:361-369 (1974).

Viadana, E., and Bross, I. D. J., Use of the medical history to predict the future occurrence of leukemias in adults. *Preventive Medicine, 3*:165-170 (1974).

Natarajan, N., and Bross, I. D. J., Preconception radiation and leukemia. *Journal of Medicine, 4*:278-281 (1973).

Bross, I. D. J., Bertell, R., and Gibson, R., Pets and adult leukemia. *American Journal of Public Health, 62*(11):1520-1531 (1972).

Bross, I. D. J., and Natarajan, N., Leukemia from low-level radiation: Identification of susceptible children. *New England Journal of Medicine, 287*:107-110 (1972).

SELECTED PAPERS IN SCIENTIFIC PRINCIPLES AND METHODS

Note: Unless otherwise noted, Dr. Bross is the sole author.

SCIENTIFIC PRINCIPLES

What is a valid scientific argument: A linguistic approach to an answer. *Methods of Information in Medicine, 16*(2):75-80 (1977).

Scientific Strategies in Human Affairs: To Tell the Truth. Exposition Press, Jericho, N.Y., 1975.

Scientific strategies in human affairs: Use of deep mathematical models. *Transactions of the New York Academy of Sciences, 34*(3): 187-199 (1972).

Design for Decision. Macmillan, New York, 1953, 276 pp.

PUBLIC HEALTH PRINCIPLES

Letter to the Editor: Possible hazards of radiologic surveillance of high risk groups. *New England Journal of Medicine, 296*(4):232-233 (1977).

Letter to the Editor: Primacy principle. *Environmental Health Perspectives,* December, pp. 329-331 (1977).

Letter to the Editor: On "Controlling one's environment." *American Journal of Public Health, 66*(9):907-908 (1976).

Letter to the Editor: Politics of mammography. *Science News, 110*(11), September (1976).

Professional ethics: Society vs. individuals. *Society for Industrial and Applied Mathematical News, 8*(3):3,8 (1975).

EPIDEMIOLOGICAL PRINCIPLES

Pertinency of an extraneous variable. *Journal of Chronic Diseases, 20*:487-495 (1967).

Spurious effects from an extraneous variable. *Journal of Chronic Diseases, 19*:637-647 (1966).

Misclassification in 2 × 2 tables. *Biometrics, 10*(4):478-486 (1954).

BIOSTATISTICAL PRINCIPLES

The role of the statistician: Scientist or shoe clerk. *American Statistician, 28*(4):126-127 (1974).

Critical levels, statistical language, and scientific inference. In *Foundations of Statistical Inference (A Symposium)* (V. P. Godambe and D. A. Sprott, eds), Holt, Rinehart, and Winston of Canada, 1971, pp. 500-519.

Statistical criticism. In *The Quantitative Analysis of Social Problems* (E. R. Tufte, ed.), Addison-Wesley Publishing Company, Mass., 1970, pp. 97-108.

Statistical criticism. *Cancer, 13*(2):394-400 (1960).

MATHEMATICAL MODELS IN CANCER RESEARCH

Blumenson, L. E., and Bross, I. D. J., Assessment of myelotoxic effects of chemotherapy from early leukopenic response: Application of a mathematical model for granulopoiesis. *Journal of Surgical Oncology, 11*:171-176 (1979).

Viadana, E., Bross, I. D. J., and Pickren, J. W., Cascade spread of blood-borne metastases in solid and non-solid cancers of man. In *Pulmonary Metastasis* (L. Weiss and H. Gilbert, eds.), G. K. Hall, Boston, 1978, Chap. 12, pp. 142-167.

Blumenson, L. E., and Bross, I. D. J., A possible mechanism for enhancement of increased production of tumor angiogenic factor. *Growth, 40*:205-209 (1976).

Blumenson, L. E., and Bross, I. D. J., Use of a mathematical model to bridge the clinic-laboratory gap: Local spread of endometrial cancer. *Journal of Theoretical Biology, 38*:397-411 (1973).

Bross, I. D. J., and Blumenson, L. E., Predictive design of experiments using deep mathematical models. *Cancer, 28*(6):1637-1646 (1971).

Blumenson, L. E., and Bross, I. D. J., A mathematical analysis of the growth and spread of breast cancer. *Biometrics, 25*(1):95-109 (1969).

Bross, I. D. J., and Blumenson, L. E., Statistical testing of a deep mathematical model for human breast cancer. *Journal of Chronic Diseases, 21*:493-506 (1968).

Bross, I. D. J., Blumenson, L. E., Slack, N. H., and Priore, R. L., A two-disease model for breast cancer. In *Prognostic Factors in Breast Cancer* (Proceedings of First Tenovus Symposium, Cardiff, 12th-14th April 1967) (A. P. M. Forrest and P. B. Kunkler, eds.), E. & S. Livingstone, Edinburgh, 1968, pp. 288-300.

SELECTED PAPERS ON PARTICULAR ENVIRONMENTAL HAZARDS

Note: Unless otherwise noted, Dr. Bross is the sole author of the paper.

DIAGNOSTIC RADIATION HAZARDS

Bross, I. D. J., Ball, M., and Falen, S., A dosage response curve for the one rad range: Adult risks from diagnostic radiation. *American Journal of Public Health, 69*(2):130-136 (1979).

Letter to the Editor: Dose-response relationship in radiogenic breast cancer. *Journal of the National Cancer Institute, 60*(4):727-728 (1978).

Bross, I. D. J., Ball, M., Rzepka, T., and Laws, R., Preliminary report on radiation and heart disease. *Journal of Medicine, 9*(1):3-15 (1978).

CIGARETTE SMOKING

Bross, I. D. J., and Coombs, J., Early onset of oral cancer among women who smoke and drink. *Journal of Oncology, 33*:136-139 (1976).

"Smoking and Health. I. Modifying the Risk for the Smoker." Less Harmful Ways of Smoking. *Proceeding of the Third World Conference on Smoking and Health* (E. L. Wynder, D. Hoffman, and G. B. Gori, eds.), U.S. Government Printing Office, Washington, D.C., vol. 1, 1975, pp. 111-118.

Sackett, D. L., Bross, I. D. J., Gibson, R. W., and Pickren, J. W., Relation between aortic atherosclerosis and the use of cigarettes and alcohol: An autopsy study. *New England Journal of Medicine, 279*(26): 1413-1420 (1968).

Bross, I. D. J., and Gibson, R., Risks of lung cancer in smokers who switch to filter cigarettes. *American Journal of Public Health, 58*(8): 1396-1403 (1968).

Moore, G. E., Bross, I. D. J., Shamberger, R., and Bock, F. G., Tar and nicotine retrieval from fifty-six brands of cigarettes. *Cancer, 20*(3): 323-332 (1967).

Wynder, E. L., Bross, I. D. J., Cornfield, J., and O'Donnell, W. E., Lung cancer in women. A study of environmental factors. *New England Journal of Medicine, 255*:1111-1121 (1956).

AUTOMOBILE ACCIDENTS

Letter to the Editor: Cars in the fuel crisis: Curb horsepower. *New York Times*, p. 42, October 2, 1973.

Highway risks at extreme speeds. *Public Health Reports, 78*(1):27-33 (1963).

How to cut the highway toll in half in the next ten years. *Public Health Reports, 75*(7):573-581 (1960).

Bross, I. D. J., and Feldman, R., Ridit analysis of automotive crash injuries. Division of Automotive Crash Injury Research, Cornell University Medical College, October 15, 1956.

OCCUPATIONAL HAZARDS

Bross, I. D. J., Viadana, E., and Houten, L., Occupational cancer in men exposed to dust and other environmental hazards. *Archives of Environmental Health, 33*(6):300-307 (1978).

Houten, L., Bross, I. D. J., Viadana, E., and Sonnesso, G., Occupational cancer in men exposed to metals. In *Advances in Experimental Medicine and Biology* (G. N. Schrauzer, ed.), Plenum Press, N.Y., 1977, vol. 91, Chap. 7, pp. 93-102.

Houten, L., Bross, I. D. J., Viadana, E., DeCoufle, P., and Stanislawczyk, K., A retrospective survey of cancer in relation to occupation. National Institute for Occupational Safety and Health Research Report, 1977. DHEW(NIOSH) Publication no. 77-178.

Viadana, E., Bross, I. D. J., and Houten, L., Cancer experience of men exposed to inhalation of chemicals or to combustion products. *Journal of Occupational Medicine, 18*(12):787-792 (1976).

INDEX